BENEATH

THE PEDESTAL

BY

Michelle Hamson

Acknowledgment

First, I thank God for His constant presence and grace.
"Each of you should use whatever gift you have received to serve others", 1 Peter 4:10

To my beloved grandmother, "Nena," whose words inspired the title *Beneath the Pedestal*, thank you for your love and wisdom from heaven.

To the women and children in shelters, foster children, and foster parents, your strength and stories inspired this book. You are a testament that love, not blood, makes a family.

My gratitude to the mental health workers, caseworkers, and all who dedicate their lives to helping others.

Thank you to my husband, kids, family, and friends for their patience and encouragement through this long journey.

And to the wonderful team at **My Book Writers**, especially Celest, thank you for your support and belief in this project.

God bless every reader of this novel.
Amen.

Dedication

The people beneath the pedestal and my grandmother inspired this novel. The names and places have been changed, and certain parts of the novel have been added to create the storyline I was looking for. I would like to dedicate this novel to my family. They heard me talk about Beneath the Pedestal for years, and little did they know they were right in the middle of it every day. It has been a long time coming, and many times, they probably thought it was not going to happen. I also dedicate this novel to all the foster children who are surviving the system, or were lucky enough to be adopted. God bless every one of them. To all those who work in the system, we know that when we wake up in the middle of the night with the worries that come with every inch of "their story," we can thank God for our love, patience, and persistence in helping these children get to the home they deserve. Whether they go back home or are adopted, a stable home and love are the answer to their prayers.

Preface

Believing someone is perfect and can do no wrong puts them on a pedestal. What is a pedestal, you ask? If you Google it, you will see that the definition of putting someone on a pedestal means that someone is greatly or uncritically admired. They cannot be criticized. Their flaws are overlooked. Have you ever had the experience of being put on a pedestal? Have you had the true love of someone who idolized you, or a parent or grandparent who could never see anything but perfection in you? Perhaps this made you feel good, but what you must know is that the positive consequences may also come with the negative. Positive outcomes can provide motivation and admiration, but they can also lead to unrealistic expectations and dependency, which can have negative consequences.

Think about your life and who has thought that much about you that they worship the ground you walk on. It can sound a little scary. In my case, it was my grandmother who told me she put me on a pedestal. I had never heard the expression before. She told me she saw no wrong in anything I did, loved me unconditionally, and I was perfect in her eyes. This did affect me throughout life, as I expected everyone to put me on a pedestal and was hurt if they didn't. It took me a while to understand my grandmother's concept, but once I did, I thought this would be an excellent name for a book. I was in high school at this time. I remember wondering what it would be like to write a book.

What type of story could I write using the name Beneath the Pedestal?

Have you ever wondered about the *people who live beneath the pedestal?* Who are they? What does living beneath entail? Would someone who lives beneath the pedestal be the direct opposite of someone put on a pedestal? These questions ran through my mind for years before I started writing this novel. There are men, women, and children beaten every day, domestic violence in homes, and drug and alcohol addiction. Some people want to commit suicide because depression has taken over their world. Some people struggle with mental health, and their families think they amount to nothing because they have made so many mistakes. Some have just been born into poverty, and they are not educated enough to know how to get out. These, my friends, are the *people who live beneath the pedestal.*

Table of Contents

The Introduction 1

Chapter 1: Hanna's Burden 3

Chapter 2: The Raid At Motel Six 13

Chapter 3: Paper Dolls And Broken Trust 25

Chapter 4: A Brand New Life 39

Chapter 5: Courtroom Chaos And Mother's Betrayal 43

Chapter 6: Behind The Parris Walls 49

Chapter 7: Safe Havens And Silent Cries 63

Chapter 8: Trisha's Downward Spiral 69

Chapter 9: The Barn Of Secrets 73

Chapter 10: Scars Of Silence 83

Chapter 11: Prayers In The Waiting Room 85

Chapter 12: Mercy Lane Revisited 91

Chapter 13: Escape From The Parris House 95

Chapter 14: Death Under The Bridge 101

Chapter 15: News Of A Mother's Death 105

Chapter 16: Fractured Justice 109

Chapter 17: The Day Of Telling 115

Chapter 18: Under Interrogation 121

Chapter 19: The Funeral 125

Chapter 20: New Homes, Old Haunts 137

Chapter 21: First-Day Tip-Off At Midville 151

Chapter 22: Adopted At Last, A Sudden Cough 159

Chapter 23: Is This The Right Direction? 166

Chapter 24: Is This The Girl I Knew? 177

Chapter 25: Ribbons And Rumblings.. 181

Chapter 26: Night Of Confessions 187

Chapter 27: A Bonfire Ain't The Only Thing Burning.............. 199

Chapter 28: Bonfire Confessions 203

Chapter 29: After The Fire 209

Chapter 30: Strays & Lines Drawn 213

Chapter 30 (B): From Caring Hands To Open Arms 223

Chapter 31: Between Rules And Desire 229

Chapter 32: The Night Behind The Bookcase............................ 237

Chapter 33: A House For Hope... 245

Chapter 34: Threads Of Faith And Practice............................... 249

Chapter 35: Paper Dolls, Quiet Scars.. 255

Chapter 36: Overture To An Ovation... 261

Chapter 37: Where's Hanna? 267

Chapter 38: Tennis Lessons, Returning Light 275

Chapter 39: Steam & Things, The Near Miss 279

Chapter 40: Between Trinity And Memory 289

Chapter 41: Winter Concerto, Hidden Crossroads 297

Chapter 42: Christmas By The River 313

The Introduction

Living beneath the pedestal, three little girls were born to an addict mother and had no choice in the cards they were dealt. Having three separate fathers, Hanna, the oldest girl of eleven, stepped in and raised her sisters. Lindsey was three and Sammi was two when their mother made a choice that would affect their lives forever.

One night, the girls are taken from the only home they know, a filthy motel room. The system places the three girls in two separate homes. Lindsey and Sami are paired together, and Hanna is placed alone.

Hanna's strength and courage prevail, as the story of these three girls comes to life. Hanna never gives up hope of finding her sisters; however, her mental health starts declining from all the trauma she endures. But Hanna has a safe place with the paper dolls she created for her and her sisters as a child. This may be her only therapy, as she fights to survive her foster homes and find her sisters.

Things are not all bad, as Hanna does have a shining Knight in Armor that brings unforgettable true love and emotional intimacy to her life. This beautiful affair of the heart happens when she least expects it, as she fights her urge to quit trusting and believing in any type of miracles that will bring her and her sisters back together.

If you believe in miracles, God works in mysterious ways, as Hanna continues to show unconditional love and support for her sisters, even when they are absent from her life. The question is, will she decide to give love a chance herself when all the cards are against her?

CHAPTER: 1

Hanna's Burden

The family I was born into consisted of my mom, my two sisters, Sami, Lindsey, and me, Hanna. We never knew our fathers, but we were told there were three of them, and each of us had our own Daddy. Our home was one of many motels, and sometimes, if my mom couldn't pay her rent, the home of her boyfriend was usually available.

Sami was two, Lindsey three, and I was eleven when our lives changed forever. When I look back, I don't know why things happened the way they did, but I do know that I was taught lessons and given strength to be able to love through it all. I believe the trauma throughout my life was only leading me to my destiny.

I remember my mom losing complete control of her meth habit when Sami was born. Sami was born addicted to drugs. She got to the point where she quit taking care of us altogether; she yelled all the time, used horrible language, stayed up all night, left us alone a lot, and slept all day. Mom used to be so organized, fun, loving, and kind, and we were her entire world and came first, but then it was like she changed overnight, and unfortunately, so did we. My sisters began withdrawing from her and came to me for everything they needed. I started to hate my mom, and I constantly planned to run away, but this life was all I knew, and I had to take care of my sisters, who else did they have?

Sami was always so sick, and due to my mom's negligence and drug habit, she inherited a feeding tube soon after birth. I remember my mom shot meth into her toes during pregnancy with Sami. I am not sure why her toes were chosen as her injection site, but the drugs took priority over taking care of herself and her kids. Mom never cared about cleaning Sami's tube or ensuring Lindsey and me were appropriately fed.

As Sami got older and began eating baby food, we discovered that she could not eat adequately and was unable to hold anything down. When she got a little older, she could eat a little, but she spent most of her days eating raw eggs and jello, which was given to her by my mom, and half the time she could not keep that down. For some reason, my mom kept up with Sami's doctor appointments. I think she did this to stay out of trouble. Once she and Sami returned home, she just went back to her old ways, and I took care of everything.

If mom had a boyfriend over and wanted to impress them, she would ensure she cooked a nice, healthy dinner that she purchased with her food stamps. But after the booze and drugs, nothing mattered to her, and I ended up cleaning up and caring for my sisters. I didn't mind; I didn't, they were so young and needed me. There was an exception; however, when mom traded me for her drug money, then there was no cleanup at all, and my sisters went to bed dirty, and I never knew who put them to bed. I would ask them the next day, but again, they were so young, I got all kinds of answers. They always seemed to be fine and untouched as far as I knew. I bet you are probably wondering how something like this could happen to children, but if you come from a background like ours, or are familiar with what goes on in the streets of drugs and poverty, you will know that this is part of daily life for many of us.

Sometimes, I would hear Mom crying in the bathroom and yelling at God. I did not know who God was. I would go looking for God in the house, but saw nobody. She would say it was God's fault and that he did not love her because he never helped her! I was eleven years old and not familiar with God or the Bible, so I just listened to her and would try to understand what she was talking about. God was such a foreign concept that I did not even question it then.

Mom told the school district I was homeschooled, but she never taught me a thing. Mom picked up second-hand books with some learning, but I taught myself almost everything. I learned a lot from TV when it worked. I knew I needed to be in school, but at this point, I figured this was my life, and I did not know anything else. I always stayed so busy cleaning, cooking, caring for the girls, and settling my mom down that there was no time for schoolwork.

It was just about every night Mom was using drugs and out meeting new boyfriends at the bar. Our motels were always upstairs in the corner, so as not to bring attention to the fact that kids lived up there and were not in school. Mom hid us the best she could, most of the time. We were told to be quiet at all times and not run in the hotel. Nobody seemed to notice because they were all on drugs, too. The Department of Child Safety and the police came a few times, but Mom talked her way out of it, I guess, because we were never taken away.

Sometimes, Mom would be in a perfect mood and tell us we were going to have fun! She would be drinking out of her wine glass and dancing to loud music around the house. Mom would even dance with us and throw us into the air, laughing all the while. My sisters would love this because she was giving them the attention that they deserved but never got. This fun would eventually wind down. Mom would get a phone call and would tell us to clean ourselves up, and we would leave. I never knew whose car we would be riding in. Going on these trips would excite us, because we never got to get out of the house to do what other kids got to do.

We did not see parks, malls, or movie theaters. We didn't get to go to gymnastics, soccer, or get ice cream with friends. We had never been in a church, and we truly did not know who God was.

We would get to go with Mom and one of her boyfriends wherever they were going, which usually was a party. We never knew what time we would be leaving, but no matter what, Mom would give us Niquel to keep us under control, and we would always fall asleep early, barely having any fun at all. I finally figured out that what she was giving us made us go to sleep, and I would pretend to take it. I would then spit it out when nobody was looking or pretend to go to the bathroom and spit it in the toilet. Most of the toilets were so dirty I hated to look at them. I would see things a child should not know while I pretended to be asleep. I wanted to sleep and not know, but I was so worried about my sisters that I just lay there and pretended to be asleep. I would finally get so tired that I could not keep my eyes open any longer, and I would go to sleep fighting it all the way.

Do your children sleep in warm beds with comforters and stuffed animals? Do they have pets like hamsters or goldfish in their rooms? I would like to know. I have never had this kind of life before. If you want to do something nice for someone, you should think twice about helping those beneath the pedestal, the less fortunate. Look, I knew at a young age that my mom wasn't a good person for what she did, and I wanted to run away. But I also knew I needed to stay and take care of my sisters. I tried to save them from the pain I knew they would start to go through as they got a little older. I had to protect them by offering myself in place of them regularly. I hope you do not want to put this book down yet. I apologize. It is just important that I get you to understand.

I have a few more questions for you. Do you realize that not everyone starts their day with goodness and thankfulness? I am sure you know that not everyone believes in God, but did you know that not everyone is told about God? Only some kids have a parent or two that is educated or have good jobs. I know it is probably hard to relate to the fact that not everyone has a car and a house and can afford groceries, isn't it?

Yep, not everyone's mom is part of a book or art club. Have you ever had your utilities shut off or had to use your clothes for diapers because there were none? Have you ever counted change so your mom could get toilet paper or cough medicine for a family member? I am sorry, but many in society have no clue about the people beneath the pedestal, the people on the other side of the tracks. They think they know of them, but they don't see the entirety of it all.

My sisters and I had a different life. Most would say it was sad, and perhaps in your eyes, horrible. There is more I want to tell you. For years, I was usually woken up late at night on these so-called fun family evenings by a stranger or mom and carried into a bedroom. I was always so tired that I did not know what was happening until I felt the presence of weight on top of me. I know this will bother you, but yes, it was to have sex with my mom's boyfriend for drug money. I bet you are wondering if I fought against it.

Of course, I did! All I can say is you quit fighting when you are told that they will kill you or mess you up very badly if you do not do what you are told. These men were not good people, and if I did not listen, they would hurt my mom as well. My mom was usually so strung out that she told me to listen because she was sick and needed their help to get her medicine, which you and I both know where her habitual drugs.

The hardest part of all of it was my mom whining about how much she loved me and how sorry she was while the sexual interaction continued to take place. All I could think of was that if I complied with everyone's requests, my sisters would be safe, and I could continue to live my life. I know this is hard for you to understand. Sometimes, I would get so tired that I would wish I were dead. Maybe you can tell me something. Do eleven-year-olds even know what death is? For some reason, I thought I knew, and perhaps I didn't. Perhaps I just knew I would run away one day with my sisters and wouldn't have to deal with my current life. Can you remember when you were eleven? I bet you had a different life than I did.

When I was often removed from my bed, taken to my mom's or my mom's boyfriend's room, and finally woke up, I would tell them whatever they wanted to hear. They sometimes would mention my sister's participation, and I would say and do anything to distract them, and I would try to distract them by telling them my sisters were too young, and it would be more fun with me because I was older. They always seemed to go for it and would leave it alone.

Knowing how to handle my life took some time to develop. I learned to organize my thoughts and be careful in planning my escape from my call to duty. My life was never anything more, and I just got used to it and became very good at pretending. As I got older, I worried more and more about my sisters being sexually abused, and I was just never going to allow that to happen. I would do anything it would take to save them. I still did not know God at this time, and I tried to handle everything on my own.

I hope you don't quit reading this. Without reading this novel, how will you ever get educated about others' real-life experiences and those who live beneath the pedestal? If you have a good, well-rounded everyday life, you probably won't deal with this often. But then again, what is everyday life? You might hear about a bit of trauma from an aunt who likes to gossip about a friend of theirs at church whose son is a drug addict and got a girl pregnant. You might see a bum on the streets and feel bad for a second, but keep driving, shaking your head in shame.

Maybe you will read about children in foster care who tried to run away and were trafficked or committed suicide because they finally gave up trying to make it alone once released into the world at eighteen. Perhaps you feel the need to provide blankets for the homeless, donate clothing for foster children in care, or make a monetary donation to support homeless children. The fact remains that society turns its back on things they don't want to be bothered with, and more people than you think don't know anything about the people beneath the pedestal.

A small percentage of people will take a foster child into their homes to love and care for them. The excuses are, that there is no time, the husband or wife will not allow it, and parents just don't want their kids to be affected by the trauma the foster children went through. Many people say, "It is just not my thing, I could never do it." It takes someone who truly wants to connect with these children, learn from them, and grow with them. It takes the right heart. Maybe that will be you? Did you know that the real world will judge me and you both? Not everyone will understand my thought process and the decisions I made throughout my life for myself and my sisters. Have you ever felt misunderstood in your life?

As far as I know, my sisters escaped the abuse. They never showed any signs, and I protected them with all my might. You might want to sit down for this next part; again, I apologize. I never thought I would tell my story to strangers. I had to perform sex acts that children never even imagined, and some adults as well. It's okay, I learned to close my eyes and imagine myself making my paper dolls, which always brought me great peace and enjoyment. They were my friends.

As far back as I can remember, I taught myself how to create paper dolls and their clothing out of paper sacks. Where did I get them? We always had paper sacks around because that's what groceries came in when we had any. I would color the clothes I would make with broken crayons given to us by the motel manager. The one I remember most was Ed. Ed would bring them over while he and my mom would discuss the motel rent in her bedroom. Mom did not want to bring me into the picture with Ed, and I don't think she liked him much. When he was around, they left me alone. I think it was because she would not take drugs around Ed, or she would get kicked out of the motel. Since she wasn't high, she had some sense, so they left me out of it.

Back to the paper dolls, I also learned to make furniture out of the paper sacks. In today's world, I guess you would think I was very talented. You have to understand this was my peace of mind, and it gave my sisters and me dolls to play with. We had no dolls; we had barely any toys to play with. My mom did not mind that I created these dolls for my sisters, as long as we did not bother her. Mom was usually high and slept for long stretches, sometimes for days at a time. When she was awake, she was usually sick.

Mom was probably at her best in the early evening. Her alcohol and drugs were actively working, but she was not drunk yet. This present moment was probably her high before her high. I always knew when it was time for Mom to go out because the loud music would start, and Mom would come out of her room wearing low-cut dresses and tight jeans, her hair and makeup looking nearly perfect. I remember never quite understanding this because she usually had her clothes on for just a few hours and was in bed with her boyfriends most of the time.

Sometimes I would get lucky, and Mom and her friends would get sidetracked drinking and doing drugs, and it would get too late, and Mom was not able to function because she was so drunk. She would just pass out. I was thrilled when this happened, because not only could I get some sleep, but I also didn't have to deal with my mom, who was now starting to notice what was happening to me. That made me feel so sad and sick to my stomach at the same time.

"Hanna, come here, baby! Mommy is going out, and I need you to keep your eyes" (as she slurred her words) "on your little sisters. Okay, baby?"

I always agreed with her, so that she would leave. "Yes, Mama, I will take care of the girls."

"That is my good little girl," she would say.

My mom would then turn off her stereo, kiss me on the forehead, and out for the night she went. I never knew what time she would get back..

Hey, I don't want you to feel sorry for me. Believe it or not, even at the age of eleven, I knew that there was a better life out there for us. However, being so young, I couldn't figure out what that was. I never saw any examples. I did meet a friend through all of our trauma, and that was Sergeant Gates. You will meet him soon. He knew it all.

I want to tell you one more thing before you read this novel, okay? Don't get caught up in trying to understand my life. Just as I am teaching you about those who live beneath the pedestal, I created my paper dolls to provide me and my sisters with what we so desperately needed throughout our life. Please note that the names of cities, schools, coffee shops, and characters have been altered. Why don't you grab your cup of coffee or tea, relax, and read a novel you will not forget.

The Raid at Motel Six

So, my friends, it was a quiet night in the city of Rutland. The Motel Six on Woodland Avenue was known for its drug lords and users, but it remained open for business no matter the circumstances. It was also a hot evening, and the motel windows were open sporadically throughout. The air conditioning barely blew out of the vents, making it miserable for the tenants, and there was the sound of fans blowing upstairs and downstairs, along with the chirping of crickets outside.

Trisha (my mom) and her boyfriend (currently) Rome, lie naked on the floor of the second-story motel. They were strung out on meth and heroin, higher than a kite. "Put your damn clothes on! You make me sick, Trisha!" The door had been kicked in, and Sergeant Gates from the Midville Police Department stood in front of the most horrific scene he had ever encountered in all his thirty years on the police force. He had many run-ins on the street with Trisha in the past regarding drugs, but was unaware she had children living in an upstairs motel. Somehow, she escaped The Department of Child Safety, and he did not know how.

Trisha could barely stand up, she and Rome lay passed out on the kitchen floor. Trisha got up, stumbled, and groaned as she fell into a kitchen chair. Gates glared at her and threw her his coat. "Put this on!" Gates continued walking through the apartment, dodging syringes and feces; the entire apartment smelled like urine. He stopped and opened the refrigerator, and saw nothing in there but a bottle of vodka and some beers. He continued through the apartment, and to his surprise, he saw a baby crib with dried feces and dirty diapers thrown all over a mattress that had no sheet. Flies covered the crib, and Gates became sick to his stomach from the smell of the room.

Gates could not help but notice the plastic tubing that was lying in the crib. "No way, please God, no!" Gates yelled to his partner, Tom. "Tom, there is a sick baby in here!" Tom came running as Gates yelled, "Tom, help me find this baby!" Both men frantically started running and looking throughout the bathroom and the cabinets. Gates was furious as he walked up to Trisha and slapped her across the face. She lay passed out in the chair. Trisha slowly opened her eyes. "Where is your baby?!" yelled Gates. Trisha looked at Gates as her eyes rolled back into her head and passed back out. "Now she passes out," Gates thought.

Gates noticed a closet in the hallway that they had not opened yet. "Tom, open that closet door in the hallway, open it", said Gates. Tom slowly and hesitantly opened the door. It was dark in the closet, but both men could see movement. Tom bent down with a flashlight, and to his amazement, he saw three little girls, one of them about two or three years old.

The girls had filthy faces, ragged clothes, and the stench was horrific. "They are alive!" Tom yelled. My sisters and I scooted into the back of the closet, and I protected my two younger siblings by throwing my arms around them." Sh Sh sh it's going to be ok girls, sh sh sh be quiet so they don't hear us." "Hi", Tom quietly said as he shone his flashlight and stared in amazement at us.

I remember being on my knees, with both of my arms tightly wrapped around my sisters to protect them. I was ready to bite these men, and I told them so! "If you hurt us, I will bite you!" I said. "I am not going to hurt you. I am here to help you," Sergeant Gates said. "I did not want them to touch us, so I spit on Deputy Tom! He brushed the spit off his face, and his arm was close to me, so I decided to bite him, and I did it hard!" "That hurt!" Tom cried out, glaring at me. "One of them bit me in the arm!" "There was no way I was going to allow these men to hurt us." "Tom, move aside, take it easy, guy," said Gates. "Gates looked reluctantly into the closet, seeing the shadow of the three of us. As he bent down," he said, "You are safe now, we are police officers, and we are here to help you. We know there is a sick baby here. Where is she, honey?"

Gates was starting to get frustrated because Trisha was not sober enough to answer his questions, and he had to rely on me for answers. Then Gates looked directly into my eyes and said, "The crib over there has plastic tubing, like maybe a baby is sick and being fed through tubes. Where is the baby?" "I had moved Sami into my lap, so I quietly pointed to her," and said, "Here she is, her name is Sami, and don't touch her!" "Can you let me see her? I won't hurt her, I need to see that she is ok, please," said Gates. "I didn't trust them, but since they were police officers and upset with my mom, I decided to listen. However, I was not going to let anything happen to us."

"Gates and Deputy Tom slowly led us out of the closet. Sami started crying because she was scared. I tried to calm Sami down as Lindsey kept running back into the closet. By the time I got them both out, I could sit down and talk with Sergeant Gates. He was very nice, and so was Deputy Tom. They explained to me that my mom and her boyfriend, Rome, were going to go downtown to the police department, and mommies were not allowed to be sleeping with no clothes on. I, of course, knew it was more than that."

"Sergeant Gates told me that the apartment was not supposed to look this way. I kind of got my feelings hurt because I had cleaned it up that day, the best way I knew how. He said that feces and dirty diapers were a public health hazard. I didn't know this; I just didn't know where to put them. They were so messy and smelled so bad. When I tried to pick them up, they would fall apart. I would try sometimes; I really would."

"My mom and Rome finally started to wake up, and Deputy Tom handcuffed them while my mom kept screaming, "You can't take my babies, you can't take my babies, I will sue you pigs, you can't walk in here and just take my kids without my consent and a warrant!". She was irate and crying. Sergeant Gates got very mad at her and said, "Your babies? Your babies, Trisha? The ones that were hiding in the closet? The ones that are filthy with matted hair that live with drug syringes and dirty diapers that are a health hazard all over the floor?" "Trisha, you have a sick baby on a feeding tube! Your kids are very thin; this is abuse, Trisha, and you are going away with your disgusting boyfriend, who, by the way, is already on his way to my squad car."

Deputy Tom had called for backup, and while the officers were watching my mom's boyfriend, he came up to help Gates with us and bring my mom down to the squad car. "Tom, you watch the girls. I'm going to be right back. I want to get these clowns to the station and get them booked." "The Department of Child Safety (DCS) will be swamped, and to get anyone over here will be a chore." We are going to have to take the girls to them." Tom nodded. Tom did not want to stay there with all the filth, but it was his job.

Sergeant Gates took my mom downstairs to his squad car and shoved her angrily into the car next to Rome, who was too strung out to care. Gates was so annoyed with both of them that he could no longer hold it in; he had been involved in my mom's episodes for many years. "You don't deserve these children, Trisha! You don't deserve all the blessings these children bring to this world." Gates got in the car with tears in his eyes and drove away as he thought, "Trisha is going to lose those girls to the foster system", but in reality, he was glad she would.

The only thing he worried about was the path we would have to take to find a better life, and we would find that life? He was a policeman and seemed pretty sure of what was about to happen. What do you think? Do you think he was worried about my sisters and me? I can tell you that he did come back for us that night, and by then, we were friends with Deputy Tom, so going down to the squad car wasn't so scary. They spoke to us on the way to The Department of Child Safety and tried to explain that they would be kind people who would help us find a home until our mom could get better. We didn't understand, but we were together, so everything was okay.

"It is a warm night, and look at those clouds moving in and here we go with more rain," Abi thought as she looked out the front window of the DCS office. The Department of Child's Safety was filled with eight kids. Between the crying, yelling, and playing, a person could barely hear a thing. Abi was new to DCS and was thrown to the wolves after two weeks of training due to a lack of employees. She had acquired over forty cases, and on this night, she was in the intake room. Not only were they short on help, but the kids were being very disruptive.

"Get back here, please; you need to change into this!" Abi said frantically. A little boy who looked about six years old had wet his pants and was running away from her. Abi raised her voice, hoping the case worker would hear her. "Do I need to give all of them new clothing?" Across the room, another caseworker raised her hand and nodded at Abi as she was trying to answer three other lines blinking at the same time.

"The front door opened to the DCS office, and Sergeant Gates walked in with all three of us. We were happy to leave Motel Six and get to ride in a police car. We hardly ever got to go outside, so this was exciting to us in an odd way. A caseworker by the name of Abi greeted us at the door as Gates whispered to her," This is a sad, disgusting, terrible case. "Here are all the details we have." Sergeant Gates handed Abi a bunch of papers. He also had some drink boxes and fruit snacks in his hands, and he gave those to us.

Sergeant Gates was an ex-Marine and held the position of a Scout Sniper. After twenty-five years in the field, he resigned and joined the police force. The stress was detrimental to his health while he was in the Marines, but he somehow maintained his character, which was that of a strong, dedicated, bull-headed, yet soft-hearted man. He loved children, and he did not see any excuse for a parent to lose a child to the system. I trusted Sergeant Gates and thought he was a nice person.

"Abi, you take care of these girls," Sergeant Gates said with tears in his eyes." I have another call coming in, and I must go now. "Sergeant Gates turned away from Abi, wiped his tears, bent down, and said, "You are safe here, Hanna, and you will be fine, I promise. I will make sure you are ok. I will check on you," He knew this would not be the last time he would see the girls.

Gates opened the front door and proceeded towards his car. He turned around, looked at the sky, and said, "God be with these precious little girls." Gates got in his car, as the other call came in again asking for his assistance. He turned on the lights and the siren and drove tirelessly away.

Abi looked sad and concerned as Sergeant Gates introduced us to her. He had said very quietly that we were filthy, hair matted, and underfed, and she needed to check our hair for lice.

"Well, hello! My name is Abi, and I work here, and I am here to help you. "I am guessing you are Hanna?" Abi squatted down in front of me. "And these are your sisters? What are their names, honey?" She was a stranger and I did not like this situation at all, but was trying to make the best of it for my sister's sake. I pointed to one, then the other. "This is Sammy, and this is Lindsey. "Wonderful!" said Abi. "Let's see what clothes we can find for you, beautiful little girls," I was mad, and tired, and sternly put my arm before my sisters, halting them from moving forward. "No way!" I said. "We don't do what you want unless you tell us what you are doing with us!" "Also, my sister needs her feeding tube!" "I am sorry", said Abi. I don't remember your names. "I just told you, and I will not tell you again unless you tell me what is happening and why Sergeant Gates says we may be going to someone else's house! My sisters are hungry!"

"Abi tried to get me to calm down. She explained that they took us from the motel because Sergeant Gates did not feel our living conditions were very good, and our mom was sick and needed help. Abi explained that we would be cared for by families that would help us until a judge could decide it was safe to return home. I was unsatisfied with this, but I could tell my sisters were getting tired, so I tried to contain myself."

"I remember looking around and seeing a room filled with kids' toys. I did not see a bed for us. We needed sleep. "Sit down here, honey," said Abi. Can you put your drink boxes and snacks down here? pointing to a table that was in the room. Let me wipe your face off and help you change your clothes. "Why do we have to change?" I asked. "Well, Hanna, it is late, and you need to put on something clean because you will soon be meeting your new foster family. The people who will take care of you until your mommy gets well. "Come on, let's find a more private room for you to change," said Abi.

"Abi led us to another room, where I helped Sami and Lindsey change. The clothes felt so good on us. They were soft to the touch and had cute little designs and people on them. Abi said they were cartoon characters, but we had never seen enough cartoons to know the characters. We were happy just to get our stinky, dirty clothes off. We needed a bath badly, and I did not understand why they did not have a bathtub. Have you needed a bath as badly as us? We never really had many at home; there was never any soap. I did try to wash us off with water."

"Abi asked our names again, and I finally told her. My sisters were getting cranky and hungry, so I asked her if we could finish the juice and fruit snacks that Sergeant Gates had given us in the car. Abi agreed and took all three of us down the hall to a room where we could be alone. Abi ventured into the office to work on something. I hugged my sisters. You guys will be okay; I will take care of you. This was no surprise to them because I was the one always taking care of them. Lindsey and Sami did not seem worried; they loved me so much that they paid no attention and trusted my judgment. I was their mother in their eyes. Did you ever have to raise your siblings? "

"Abi came back to get us and took us into the playroom. Lindsey and Sami were tired but had never seen so many toys, so they played with everything! Abi took me aside. "Hanna, I don't want to upset you, said Abi. I know you are very close to your sisters. I need to tell you something; you will have to spend the night in a different house from your sisters just for tonight." "What?" I said. "No! I don't leave my sisters." "You see, Hanna, there is a problem with the available foster families' bedrooms. There are two families; one has two beds in a room, and the other has one bed. Since the two little ones are so close in age, we thought they should stay together. Hanna, it is just for one or two nights until another foster home has beds for the three of you."

"I was furious at Abi, and I let her know! "I told you no! I will not leave my sisters!" "Hanna, I will ensure Sami gets a feeding tube at the foster home tonight." "No! You get one now! My sister will die without her feeding tube! Abi quietly brought me into the office." "Look, Hanna, I have read the report and know we must do this immediately." "I had tears in my eyes and a lump in my throat, but I nodded in agreement. Tonight, Abi." "I know Hanna. I am working on it, and I can promise she will have it tonight before she goes to bed."

"We all finished our animal crackers and juice, except Sami. Sami never ate well, so that is why a feeding tube was needed. Abi had all three of us continue to play in the playroom. I noticed a dirty couch with crumbs all over it. I noticed it because it reminded me of the motels we called home."

Abi returned to the playroom a few hours later. My sisters were so tired that they fell asleep on the couch. I had brushed all the crumbs off the couch they lay on, so they would be more comfortable. Abi woke the girls up and had all of us follow her down the hallway into a large conference room. There stood two adults who were unknown to us. My sisters snuggled up against me for security. "Girls", Abi said. "This is Ms. Walker and Mr. Glasso they are your friends and are here to help you." "They will be taking you to your new foster homes.' "A foster home is a home with foster parents, and you live there until mommy can get well." "By this point, I was not happy at all." "Why are there two of you?" "Well, Hanna," said Abi; there are only two beds at the home Ms. Walker is taking Sami and Lindsey to. We have to separate you for just a little while, Hanna." "No! I yelled, "My sisters can't be without me!" "Hanna, both homes are great, and we will ensure you see both of your sisters as much as possible, I promise", said Abi.

I put my head down and began to cry." "Please, Ms. Abi, please don't separate us." "Hanna, we have no choice; there are simply not enough beds in the available foster homes to hold three of you tonight. We know that Sami needs medical attention. Ms. Walker has a feeding tube and will ensure she gets it tonight. Training will take place tomorrow morning at the foster parents' house. If you understand me, Hanna, the people watching Sami and Lindsey will get training on the feeding tube and what Sami needs to be healthy." "All I could do was agree with a nod as tears rolled down my cheeks." "I had never felt so broken or so sad. We were better off in our dirty motel; at least we were together."

"It is getting late; we have to get you girls to bed," said Abi. Lindsey and Sami were exhausted and walked around holding my hands, barely standing. I hugged them tightly and promised they would go to my friend's house for just a little bit, and I would be there in a little while. It was the best I could do for them and myself. Ms. Walker took the hands of Sami and Lindsey as I slowly released their hands from mine. All I remember is their sad, dirty, and tired faces as they turned to look at me one last time, as they went out the front door with someone I knew nothing about. I yelled to them, "I will call you tonight, and I love you!" Abi quickly stood in front of me. "No, Hanna, there will be no calls while the girls are in foster care for a few days. The visitations between you girls need to be set up. You will need to be patient."

I was very, very upset and could barely think. If this were you, how would you respond? "Abi, you said that I would see them!" "Hanna, I am tired, and I know you are too," said Abi. 'You will go with Mr. Glasso to your foster parents' home bright and early tomorrow. We must keep you on the couch tonight since it is so late. I will be staying with you. Your placement has requested you come in the morning instead of tonight since it is so late."

Abi made a bed for me on the couch, and I was so tired I fell asleep immediately. I am sure that may seem odd to you. It is not that I did not care about what was going on, but my hands were tied, and I was young. What was I to do? I had nobody to speak up for my sisters and me but me.

"Abi went back to her office and fell into her chair with exhaustion. She felt sad and guilty about lying to us. She knew it would be a long while until we would go home, and she also knew it would be a long time before we would see each other due to the location of the foster homes, because they were located in two different counties."

CHAPTER: 3

Paper Dolls and Broken Trust

Mr. Glasso was a witty man, very impatient, but had a heart of gold. Abi woke me up and gave me a granola bar and an apple. She explained Mr. Glasso was waiting for me, and I needed to hurry and get my shoes on. Mr. Glasso was a caseworker for Social Services. All I had were my pajamas that they gave me and my old, dirty white tennis shoes. "Hanna, you will be fine. You go with Mr. Glasso, and he will take care of you and get you to your new foster home with some nice people who will take care of you. Sami and Lindsey are doing just fine," said Abi.

I did not know what to believe, so I listened and went with Mr. Glasso. Mr. Glasso opened the car door, smiling at me like I was supposed to know him and be happy to go for the ride. I got in, and he leaned over to seatbelt me. I pushed away his arm and buckled myself in. Mr. Glasso looked overly worked and stressed already early in the day. I left it alone and just stared out the window, hoping my sisters were doing okay and that they were safe.

It seemed we were driving forever, and I was getting tired of Mr. Glasso's jokes. The road had turned to dirt and now was staying that way. "You will like these guys," said Mr.

Glasso. "They are good people, and they know how to farm. All kids should learn how to farm." I just kept looking out the window, staring ahead, wishing we could get this over with. I was so worried about Sami and Lindsey, and confused about how everything was moving so fast. "Mr. Glasso, I need to pee." "You need to go to the restroom is a better choice of words, Hanna," said Mr. Glasso.

Mr. Glasso seemed proud of his 1999 Toyota Corolla. He kept it spotless inside and had new seat covers; at least they looked new. I found out it was a company vehicle, so that is why he kept it so clean. Mr. Glasso pulled up in front of a white country house, and way down the other end of the property was a red barn. I noticed there were not many other houses around. "Well, this is it," he said. I was used to motels and not getting out a lot, so this white house was beautiful to me. I found out later that it was a three-bedroom, two-bath house over 2500 square feet. This was huge!

A woman with long, straight brown hair with streaks of gray walked out on the porch and motioned Mr. Glasso to approach the house. Two spirited German Shepherds barked and ran up to Mr. Glasso's car. "Get away now, get away," Mr. Glasso said as he opened his car door and motioned the dogs to get away. The dogs moved away from the car door, sniffed Mr. Glasso, and returned to the porch. I was sitting in the back seat, belted in, and wishing everything happening to me would be over. For a moment, my mind drifted off, and I thought about all the paper dolls and clothes I had created for my sisters and me. I wondered if they still had them in the plastic bags I had given them before we left home. I was hoping they could play with them at their new foster home.

"Hanna, we are here," Mr. Glasso sighed. I was not happy about it, but I tried to be halfway interested and brought myself back to reality. "Get your things out of the back seat," said Glasso. I did not have much besides the P.J.s Abi gave us at the DCS office. I had a plastic bag of torn paper bags I collected to make paper dolls for my sisters and me. I also had my oversized slippers, decorated with sequins, but they had holes in the bottom of them. I did not have anything else, and I think Mr. Glasso was thinking of some other kid, because he did not even put anything in the back seat for me to do or to possibly wear. Have you ever had to travel to somebody else's house, and you did not have anything to bring?

Mr. Glasso opened the vehicle's back door and motioned me out of the backseat. I slowly put both legs in front of me, sliding out of the car. I stood before Mr. Glasso with a blank look and empty sadness. I did not want to be there. As bad as things were at home, I just wanted to go home; it was better than this, and at least I had my sisters at home. Mr. Glasso took my hand and asked if I had everything I needed. Of course, I did not answer and continued to stare into space with an empty look. And then, he did something terrible! He noticed the plastic bag that I brought from home. He grabbed it and started crunching it up, thinking it was trash. "Hanna, I told you not to leave trash in my car. It is a company car, and I must take care of it." Mr. Glasso handed Hanna the crunched-up bag and pointed to the barn. "Hanna, please take this trash bag to the trash can next to that brown truck over there."

I felt my heart start pounding and the blood rush to my head. I began to shake and sweat uncontrollably, and my mouth became dry. I knew I had to make this scumbag of a person understand that this was my only connection to my sisters! That bag that held my paper dolls was the only happiness, laughter, and connection that I had with my sisters, and nobody was going to throw away that bag!

My sisters loved the dolls I made, and they were starting to learn to make them themselves. Nobody was going to take the paper dolls away, nobody! Nobody was going to take my paper sacks! I did not see Mr. Glasso as a person trying to help me anymore. I was too angry, and all I saw was a thief who was trying to steal our peace. Have you ever been so mad at someone that you felt like you had to stop them, and you would do anything to do just that? So then I lost it! I could not stand the pressure anymore and lunged at Glasso, punching, scratching, and kicking him, trying to save the bag from his destruction! I screamed at the top of my lungs! "Leave it alone, you stupid man! Leave my dolls alone!" I was not going to allow this stranger to take away the only fun and peace we ever had together.

Mr. Glasso restrained me and told me to calm down. "Hanna! What is going on here? What is in the bag? Let me look Hanna!" Mr. Glasso looked into the plastic bag and saw nothing but a brown paper bag. "Hanna, this is trash, honey." I was so angry, and I started swinging my arms as to hit Mr. Glasso. "Leave it alone, you stupid," I said.

Glasso quickly gave me back the bag. "I am not sure what this is; apparently, it is something to you. But, you continue to act violently, and you will be transferred to a home that will allow minimal privileges and no visitations with your sisters!" said Glasso. I did not care to explain myself, I just wanted my plastic bag with my scraps of paper sacks so I could continue making dolls for my sisters and me. I got back in the car and sat down in annoyance.

It took a while, but Glasso finally calmed down. I guess he understood children in poverty hung onto certain things as security because their lives were so unstable. "Now, let's go, Hanna," Glasso sighed, and looked down the property to the porch where Mrs. Parris was standing. He knew she witnessed my outbreak of aggression, and he was hoping that she did not ask him to remove me before they even had a chance to get to know me.

Glasso and I walked towards the porch and were soon standing in front of Mrs. Parris. She was a quiet woman and very beautiful. She and her husband were experienced foster parents of two years and had fostered quite a few children during that time frame. "Looks like there was a problem out there?" Mr. Parris opened the screen door and invited Mr. Glasso and me in. The Parris's had lost their only daughter due to drug usage and a fatal car accident just over two years ago, so Mr. Parris had felt foster care might help them get over the pain. They both agreed that devoting their time to children in need would help them overcome the death of their sixteen-year-old daughter. Fostering young girls seemed to fill the void for both of them.

"Would you all like something to drink?" asked Mrs. Parris. "I am fine, thank you," said Mr. Glasso, as he nodded at me to let me know it was ok to accept a drink. I felt very empty and sad, yet determined. I looked at Mrs. Parris and sternly said, "No, I want my sisters. You think because I am at a different house and around different people, I don't want my sisters?" I began to cry, and it turned into uncontrollable sobbing. "They told me that we would be together!" I yelled at the top of my lungs. Mr. Glasso told me to calm down.

I looked at Mrs. Parris for immediate answers. "Can you find my sisters?" Mrs. Parris took my hands into her own, bent down, and looked into my eyes, which were filled with sadness. "Hanna, I know you are hurting. I know you want your sisters with you, and you are scared. I am not sure of your future, but I can tell you that you are safe here and that we will support whatever visitations Mr. Glasso sets up for you. Please, honey, come this way and let me show you your room. We just painted it and added daisy décor, we think you will like it here."

Mr. Glasso whispered, "Hanna, here is my card. If you need anything, I'm a phone call away." Mr. Glasso smiled at Mrs. Parris and signaled me with a wave of the hand to go with her to see the new room she was so proud of. Mr. Glasso waved goodbye to me and said, "See you soon." I felt emotionally numb, but followed Mrs. Parris down the hall and into my presumably new bedroom. It was freshly painted, as Mrs. Parris said it was.

The room had hand-painted dressers with stenciled daisies, and a clear glass vase of fresh daisies was centered on a beautiful pink lace doily. As tired as I was, I thought it was beautiful. "I hope you like daisies, little lady," said Mrs. Parris. I continued to look around the room and saw that it was painted yellow, with a large white daisy in the middle of the wall. The bed had a white bedspread. I had never seen anything so beautiful before and never had a dresser or a bedspread, so I thought, "I can handle this until I can see my sisters."

"Well," said Mrs. Parris, "I know you are tired and we have a lot to talk about, but I will leave you here for a bit to get to know your place here. Those boxes on the end of the bed are all yours, Hanna. You will see some items to make arts and crafts, and maybe if you like to string some beads, they are in the mix as well. I thought they would keep you busy after school and on weekends." Mrs. Parris left the room.

I stood in the entryway, marveling at what I saw. I did not have any clothes or anything to put away in the pretty stenciled dresser, except my bag of paper dolls. I decided to put them in the bottom drawer. I looked through the boxes on the bed and decided to move them onto the floor because I wanted to go to sleep and forget this was even happening to me.

As I laid the boxes to the side of the bed, one by one, I noticed a little black rolly polly bug crawling on the carpeted floor. I picked it up and rolled it in my hand over and over again. I walked over to the window, still carrying the little black rolly polly bug, and looked outside. Everything looked miles and miles away. Away from what? I thought. My home is in a motel. It was so green here and seemed so quiet. The trees and grass were bright green.

For a second, my thoughts got lost, and I felt guilty that I was not thinking of my sisters. I put the rolly polly bug back down on the floor and came back to reality. I sat down on the bed and thought, "I don't want to be here, and I will find you, and when I do, we will never be apart again!" I lay down on the bed and started crying, and must have fallen asleep when I woke up startled and heard Mrs. Parris yell, "Hanna! Come on out for dinner!"

I was hungry, exhausted, and pretty much done with the foster placement ordeal. "I cooked some rice and baked chicken with just a tad of mushroom sauce," said Mrs. Parris. She continued speaking loudly as if I was not listening. I just continued to lie on the bed and take deep breaths so I would settle down. Mrs. Parris continued repeating herself. I finally got so sick of hearing her that I got up and walked towards the door. "Hanna! Do you hear me, child?" Mrs. Parris was getting frustrated with no response from me, and there was no sympathy in the cards as she kept yelling.

I turned around slowly in the doorway and looked back at the window in my room. The trees were blowing, and dark clouds were moving in. I thought about my sisters and continued worrying about Sami's feeding tube. It was dinner time. Did anyone take the time to clean it properly or at all?

"What are you doing, Hanna?" Mrs. Parris tapped me on the shoulder and locked arms with me, and brought me into the kitchen. I slowly looked around, feeling nauseated and exhausted, and even though I was hungry, I did not feel like eating. "Look, Hanna, it is late, but I want you to have a good dinner before you go to bed." Mrs. Parris looked at me with a skeptical kind of grin. "Thank you," I whispered. "Baked chicken and rice, yum yum, you like this kind of stuff, girl?" I nodded just to appease her as she pulled out a kitchen chair for me to sit on.

I slowly began to sit down when I felt the chair being pulled out from under me! I stumbled so as not to fall. "I am sorry, girl, but nobody sits at my table without washing their hands!" "Mrs. Parris, I was going to, you did not give me time." "I know you were not going to, because I did not hear the water turn on in the back of the house!" I got up off the floor and stood upright in front of Mrs. Parris. I was not scared of her, I had been through so much in my life that this was just another everyday thing. I was not expecting it and was a bit surprised though. "Look, Mrs. Parris, I am going to wash my hands right now." Mrs. Parris nodded her head in agreement.

I walked into my room and to the bathroom, washed my hands, and came back out to the kitchen. "Yes?" said Mrs. Parris. "I washed my hands, ma'am." I continued to stand right in front of Mrs. Parris, looking her in the eyes. "So I gave your dinner to the dogs; they appreciate being fed." Mrs. Parris turned around to the sink and continued doing the dishes. "It's getting late, Hanna, you need to go to bed." "I don't care," I thought as I walked down the hallway to my room. Mrs. Parris never said goodnight, she never looked up, and she never apologized for allowing me to fall on the floor.

Once back in my room, I saw some clean towels on my dresser. The shower did sound good, and I wanted to wash my hair. There was shampoo, conditioner, and a bar of soap in the shower. I got in, and the water felt so good that I almost fell asleep. It had been days since I had a shower, and as I started wondering if my sisters were having one, I heard someone in my room, and it sounded like they were going through my dresser drawers.

They were opening them and shutting them like they were looking for something. I did not want to get out of the shower because I did not trust Mrs. Parris, and I did not know who else was in the house. I waited a few minutes, and I saw a shadow through the shower door move quickly past the bathroom entryway. The shower door was made of frosty glass, and whoever walked by the door could see me in the shower as well, but not a clear picture of my naked body.

I slowly got out of the shower, grabbed a towel, wrapped it around my body, and walked into the bedroom. The bedroom door swung open, and there stood a man smiling at me. "Hello, young lady. My name is Mr. Parris." I felt so uncomfortable and turned around and ran back into the bathroom shutting the door. You may ask yourself how a girl who was constantly sold for drugs could feel uncomfortable, but you need to know that those situations were what I was put in and they were not something I wanted to be doing.

Mr. Parris started knocking on the door and laughing. "It is alright, youngin, I am not going to hurt you. Come on out, and I will show you where your nightgown is." I locked the bathroom door and stood there with tears rolling down my face. Mr. Parris continued to knock. All of a sudden, I heard the bedroom door open and heard Mrs. Parris. "Are you going to come eat or what? I am sitting out here by myself, and here you are playing knock-knock with our new foster daughter!" "I am coming, Mama. I was just trying to show Hanna where her jammies are."

Mr. Parris continued to laugh, and they both left the room. I heard them arguing in the kitchen, so I quickly started looking through the drawers and did find a nightgown with some farm animals on it. I quickly took it back into the bathroom and put it on. I wanted to call Mr. Glassco, but these people did not do anything I could show proof of yet. I could not just call him and complain, and I had no proof. I was just a foster child, so most likely he would not believe me.

I checked the bedroom door and locked it from the inside, hoping they would not try to get into the room as I slept. Now I was exhausted. I just wanted to lie down and plan how I was going to see my sisters. My bag of paper dolls came to my mind. It was the only thing that gave me comfort and would make me feel close to Lindsey and Sami.

I walked over to the stenciled dresser and pulled out the paper bag with the paper dolls from the bottom drawer. "I wonder if Mr. Parris saw my dolls," I thought. If it was him that entered the room, he went through the drawers so fast there was no way for him to observe in detail what he saw, right? It just kept running through my mind as I laid all the paper dolls out on the bed. I started making a dress for one of my dolls and soon fell asleep with them all over my bed. I slept through the night and woke up with the light still on and a mess of paper dolls all over my bed. It seemed as though nobody came to check on me, and that was just fine with me.

It was morning, and I got up and started picking up my paper dolls quickly in case someone would see them. I put them back in the middle drawer instead of the bottom drawer, in case someone had seen them in the bottom drawer and tried to see them in the future. Then I thought that was stupid, and removed them and stuffed them below the sink behind a bunch of towels, hoping nobody would find them until I could find a better hiding place for them. My paper dolls were mine, and nobody needed to know about the developing relationships I had with my dolls, especially the one with Nadine; she was always so evil and hard to deal with.

As I stood in the bedroom and was trying to decide if I should leave my nightgown on or put on my old clothes, the door opened. "Hello, Hanna, we meet again." Mr. Parris had used a key and unlocked the door. I felt my stomach go up through my throat. "Hey, young lady, I heard that you are ready to play some fast paced racquetball today?" "Sir, I don't want to play racquetball, I have never played before and I have not eaten as of yet." "Well, we are going to go out and have some fine breakfast first and then we will go play, sound good?" "No sir, I don't want to play. I am just not up to it right now." Mr. Parris threw a plastic bag with a bunch of clothes in it to Hanna. "Listen, you put some of these clothes on, something for the heat, and we will eat and get on the road!" Mr. Parris left the room smiling and laughing. I felt sick to my stomach but managed to find an outfit, put it on, and head into the kitchen for breakfast.

Mrs. Parris smiled and acted like nothing had ever happened, asked me how I slept, and served me pancakes, eggs, and sausage, which I scarfed down without a problem. "Do you want some milk, Hanna?" Mrs. Parris asked. "Yes, ma'am," I said skeptically, thinking she may throw it at me. Mrs. Parris put it in front of me, kissed her husband on the cheek, served him coffee, and walked out of the room without a word.

Being left in the kitchen with Mr. Parris by myself, I began to speak nervously. We talked about pancakes and syrup, and how hot it was outside. I felt Mr. Parris was beginning to be a little easier to talk to than Mrs. Parris, so I decided to mention his coming into my room unannounced with a key. "Sir, would it be ok if you did not just enter my room without me knowing about it? You used a key to get in, and I was not decent." Mr. Parris did not look surprised and answered in a very calm manner. "Hanna, I am sorry, hun, I am just used to this being my house and coming and going like I always have without many people living here. We do have foster children periodically, but it has been a while. I will give you my word that it will not happen again."

"Thank you, sir," I said, trying to hide my trembling hands. I did not believe him, but it was appeasing him for the time being and hopefully delaying racquetball.

I finished my breakfast as Mr. Parris watched me eat every bite. "Ok doll, are you ready to go play tennis?" "Mr. Parris, please sir, I am just not feeling that good. Please don't make me go play today." "Hanna, it will be good for you. Follow me into my office. I bought you a new outfit to wear today." "I thought I was wearing what you gave me out of that bag of clothes you gave me?" "Nope," said Mr. Parris, "come here." I followed him into his office, not knowing what was going to happen to me and fearing some of the thoughts that came into my head.

CHAPTER: 4

A Brand New Life

Girls, I asked you to get your white leotards from the bottom drawer, and quit running in the house!" Lindsey and Sami were running about playfully through the house that they now called home. The couple who took them in on the night all the girls were taken to DCS was a Christian couple who could not have children. Emma and Canby Black felt that their calling was to adopt a few children from foster care. They did not want infants since their clock was ticking and they would soon be in their mid-fifties. When introduced, they bonded with the girls immediately.

"Girls! Come here! Look at the mess you made in the closet. We were going to go through these shoes and see what fits. Now they're everywhere." Emma tried not to laugh; she thought these girls could do no wrong, and they were absolutely beautiful to her! She loved how close the girls were to each other, and they would mention their sister, but they were young and at this moment in time, toys and love were what seemed to keep them truly happy. The Blacks had received many shoes from other foster parents, and they wanted to try them on the girls to see what fit and what they could give to others. The girls had seen all the shoes and decided to play a matching game that created a small catastrophe.

Sami and Lindsey continued running through the house giggling, and suddenly ran over to Emma and stopped with a screeching halt. "Sami hit me," said Lindsey. "Lindsey, I am sure you can handle this; you are older than Sami." "No, she hit me because I told her to keep her shoes on, I don't want her to catch a cold!" Sami said. "This is no reason to hit." Emma smiled with a warm, loving smile and motioned the girls to come with her as they skipped over to the closet. "Look, guys, we need to pick this up," as she pointed to the shoes all over the inside of the closet. The girls sighed and threw their little fit, whining, but started picking up the shoes as they were told.

They began to sing a little song. "Cleaning up is lots of fun, because when I tickle you... you will run, cleaning up is lots of fun, we are almost done, so run, run, run!" The girls would giggle and giggle and jump up and down, tickling one another while singing this song. They looked so happy while singing. "Where did you learn that song?" "Did you just make it up?" "No, my sister Hanna taught us," said Lindsey. Sami put her lower lip out and said, "My sissy, my sissy," and started to cry. Emma felt the tears building up in her eyes. "You know, I think we could use a great big ice cream cone after this, what do you think, girls?" "Sami can't have ice cream unless her tube is cleaned first." Lindsey was stepping up to the plate since Hanna was not around and was showing her concern for Sami's well-being. "I know, little one," said Emma as she poked Lindsey in the belly. "You are the best big sister ever, and you, my love, will get a huge bubblegum-pink ice-cream cone with your baby sister today! But you gotta hurry, hurry, hurry to get all the shoes picked up so we can go!" The girls both smiled, and Emma closed the lid on the shoe container, pushed it back into the closet, and took Sami's hand as she explained to Sami that she was going to check her feeding tube. Sami agreed and all three of them ventured into the bathroom to check the tube and get cleaned up for ice cream.

The ice cream parlor was packed, causing Emma and the girls to stand in line. Lindsey looked out the front window of the parlor as her mind began to wander off. She wondered where her sister was. She missed her terribly. She knew that Hanna was always ok. She was their rock, and Lindsey knew that she would someday see her sister again, because Hanna would make sure of it. Hanna would find a way for all three of the girls to live together again and be happy.

Lindsey felt content and safe living with the Blacks. They were kind and made sure they were taken care of. They watched Sami's tube like a hawk and made sure it was always clean.

Lindsey knew her mom had never done that. The girls had plenty of clothes and toys, and lots and lots of food. As time went on, they had friends from playgroups and church get-togethers and even got to know the kids next door who were close to their ages.

Life with the Blacks was everything Lindsey and Sami had once only dreamed of. Lindsey missed their mom, but living with her was so different from this life. Lindsey knew her mom was always crying and talking funny, and besides, Sami seemed healthier and very, very happy and was talking so much more!

"Don't drip your ice cream down your chin, little one," said Emma as Sami tried to eat it all at once. Sami smiled and took her sleeve, and wiped the ice cream off her chin.

Emma was in awe of how beautiful the two girls were and knew she was doing something right in her life by fostering them and hopefully adopting them. She continued to daydream as the girls finished their ice cream, but the innocent daydream soon turned into a shattered memory of the past. "Mommy, mommy," said Lindsey, "you spilled your drink all over the table." Foster children always tend to call their foster parents mommy and daddy, and Emma was fine with this. She jumped up and grabbed the napkins to wipe the spill up. "It's ok, girls, I just am so clumsy sometimes, really, it's okay, I got this." Emma wiped up the table and smiled at the girls, trying her hardest to avoid the memory of her father. The memory that she could not forget had left her without being able to have children.

Courtroom Chaos and Mother's Betrayal

Having three girls never felt like a burden to Trisha. Staying higher than a kite and sleeping with those who supported her addiction was just a daily routine, and as long as the girls were kept quiet, she stayed out of trouble. Trisha would sedate the girls with Niquel before she would ever bring anyone over to the hotel they lived in. If at a party and she wanted to stay, she would have someone watch the girls in one of the bedrooms, and still would make sure they were under the influence of Niquel.

"I don't want to take that nasty stuff, Mama," said Hanna. "Hanna, just take your night-night medicine. You will have dreams of beautiful fairies and princesses, no other night-night medicine does this," Trisha would say. Hanna and her little sisters took their night-night medicine and quickly fell asleep as one of Trisha's girlfriends sat in a recliner reading a book. "You don't have to stay here long," said Trisha. "Once they're asleep, you can leave them here and join us in the living room. We've got some partying to do!"

Many times Hanna was the subject of sex for drugs since she was older. "It's no big deal," Trisha would say, "It will be over before you know it, and we'll all get what we want, baby." "You, my beautiful baby, will get whatever you want in life. Be the best you can in everything you do, Hanna. Anything, anything at all," she would say in her drunken stupor. Trisha knew that Hanna was being hurt emotionally, but it was the drugs that did the talking, and she would not give in to the thought of sobriety since that was just too hard to even imagine.

"Give me a cig, hun," Rome said. Rome and Trisha were both waiting impatiently outside the Midville Courthouse, smoking one cigarette after another. Today was the day that Trisha was to find out what the judge had to say about her recent arrest and losing the girls to the system.

"Quit smoking in my face," Trisha snapped, "It is making me cough and smell." Rome chuckled as he gazed towards the mountains, thinking he could not wait to get back to the trailer and have a beer. "Man, I sure am hungry. Mmm mmm mmm. I sure miss Hanna making us some good food, and giving me that good stuff, mmm mmm mmm." "Shut up, Rome," Trisha said, "and you better not let on about anything! I want my babies back," she said as she began to cry. "Just a faker you are," said Rome. "You know those girls mean nothing to you."

"They are your ticket for the yummies and me, baby." Rome put his arms around Trisha. "MMM MMM MMM, why don't we get out of here for about 15 minutes?" Rome chuckled. Trisha stiffened and pulled away. "Let's go make ourselves feel good, baby," said Rome as he lit up another cigarette. "Shut up!" Trisha said, kicking the gravel in the parking lot.

"Hey, do you know what time it is?" Trisha said as she stopped a well-dressed woman going into the courthouse.

"Three twenty-two," said the woman. "Shit!" "We gotta get inside," Trisha looked at the woman and gave her a thankful nod. "Come on, Rome, we are gonna be late!" Trisha grabbed Rome's hand, and they hurried into the courthouse. They had to go through a metal detector and an X-ray machine operated by security personnel. They were required to empty their pockets as well. Trisha was a lot of things, but not a thief. Rome, on the other hand, was arrested for anything and everything.

"Hey, babe, you go on, I am hungry and I just don't think I can sit a long time in there." Rome was impatient and did not want to be bothered by the girl's case at all. Rome did not care about Trisha and her girls; he just used them for a place to stay when he was too drunk to drive back to his trailer and sex with Trisha. Trisha rolled her eyes, "Okay okay," she said, as she motioned Rome to go ahead and go.

Trisha walked haltingly into the court hallway. "Where do I go?" The lady behind the counter looked at Trisha and asked, "What is your last name?" "My God, you people should know who I am by now. I have been here numerous times and you waited on me, said Trisha!" "I need your last name, ma'am, to look up what court you need to go to," this time the lady said it sarcastically. Trisha rolled her eyes. "Trisha Flemming, you witch," the lady muttered as she turned and walked away. "Hey, you bitch, I'm talking to you." The lady walked out of the room as Trisha continued to yell.

A tall, thin woman with dark hair followed the clerk out of the backroom. "Hello, Trisha. I am the supervisor for the clerks at this courthouse. In fact, I believe I have met you before. Due to your rudeness to my staff, I would like to ask you to step outside and adjust your attitude before you come back to the court." "Look, I am sorry, I am trying to get my kids back, and this lady was irritating me. I am sorry, ok? I will be good and go in and sit down if you tell me which courtroom," said Trisha. "Trisha? I told you to step outside for a bit and adjust your attitude." "I will be late." Well, you should not have treated my clerk the way you did. Step outside, please." "Whatever!" Said Trisha angrily.

Trisha walked aggressively over to the stairway and went downstairs to find Rome. She could not see him, so she smoked a cigarette, went back into the court, through the metal detector and X-ray machine, and emptied her pockets once again. Trisha took the elevator upstairs and walked over to the same clerk she seemed to be having problems with. "Ok? Can I go in now? I adjusted my attitude." The clerk looked at Trisha and said, "Let me get my supervisor."

Trisha waited about five minutes and was just about to stir up some more trouble when the supervisor came out and said, "Trisha, you may go into Courtroom number two." Trisha opened the door and walked in, and sat down in the first row. She knew from past experiences that rowing was the easiest way to get to the stand faster. There were not as many people seated as she had seen in the past. She decided to move to the third row. "All rise." Everyone stood up as Judge Arias walked into the courtroom. Trisha hated Judge Arias, probably because he knew her very well and was very tired of her appearances in court with no follow-through or improvement.

"Please be seated," said Judge Arias. Judge Arias was a nice-looking older man with grey hair, probably about sixty-two. He had been around a long time and had been a foster child himself. Drugs and child abuse were his area of expertise. He was definitely in favor of all children.

"Hello, Trisha. I see you are back at your home away from home," the judge said sarcastically. "Judge Arias, I have really been cleaning up." Trisha was still in the third row and decided to give her opinion even though she was not called on. "You see, the cops came and busted me, and they really made a big mistake because." "Trisha Flemming, please be quiet." Judge Arias was fed up with Trisha and her ways and did not feel like listening to her lies again. Trisha was representing herself and decided to quiet down. "Trisha, stand up." "This courtroom does not need another liar." "Please proceed to the stand." Trisha stood up and walked over to the stand, and sat down.

"Trisha Fleming, since I met you three years ago, you have been in and out of jail, recently lost your children to the system, and have failed to stick to any plan of action to even help your family. You're a repeat drug offender, never holding a steady job, and never creating a stable home for your girls. You're wasting our time." Judge Arias slammed his hammer down on the desk. "What in God's name is different in this case from the others? Trisha, you lost your children! This time, they are in foster care, Trisha. You worked hard to keep all your little secrets from us. It worked for a while, but they have all come to the surface now," Judge Arias said, frustration heavy in his voice. "Deputy Tom, take Trisha out of this courtroom now. She can reschedule a court date when I have calmed down." Deputy Tom walked over to Trisha and motioned for her to head toward the door. Trisha looked at Judge Arias and started crying. In her raspy voice, she said, "You are a mean man. My girls need me, and you will pay for this." Judge

Arias got up and walked out of the courtroom, not having a second thought about Trisha Flemming. He was done with that today.

CHAPTER: 6

Behind the Parris Walls

Hanna lay on her bed, exhausted, staring out the window, thinking about her sisters and wondering how they were and what they were doing. Mr. Parris had given her a new dress to wear to play tennis and allowed her to change into it in her own bathroom. They did go play tennis, and nothing happened as she expected it would, so she let her guard down and decided to trust him, at least a little more than Mrs. Parris. She did not feel she could tell him about the chair incident. She also thought about her mom, but felt she was dealt a bad hand when it came to her family, and none of her mother's choices were her fault. She knew there had to be a better life, and she was determined to find it for her sisters and herself.

Hanna knew she needed to talk to her caseworker about getting out of the Parris's house, but she wasn't sure how to begin. She also knew that if she could get someone to listen to her, she could request her sisters, and they could all live together.

The door creaked open slowly, and I caught a glimpse of Mr. Paris smiling from the corner of my eye. He threw open the door and said, "Peek a Boo!" I just turned my head and tried to ignore him by looking out the window, pretending not to hear him. "Hey girl, are you ready to play some more tennis with me today? Do you have your cute little dress from yesterday that you can put on?

"It is good exercise and it will keep that little body of yours in shape. Now go get the dress, Hanna."

"No, thank you," I said as I continued to look out the window.

Mr. Parris gave Hanna the creeps and there were times she felt someone watching her at night and when she finally did start to wake up, it was like someone ran out of the room quickly. "Hanna, you need to get your exercise."

"I just played last night, sir. I just got here, and I just would like to lie on my bed and relax, and I don't want to play."

"Well, ok then, I will make arrangements for an early evening time with my wife, ok?"

"Your wife will like that," Hanna said.

"No, I will talk to her about you going with me later this evening and make sure we can eat dinner first. We don't want to upset the boss."

"I told you I don't feel like going today."

"You listen up, Hanna, we took you on here at this house to help you. We help others, and your caseworker will be happy to know we are doing just that. My wife will let you know what time you should be ready for later, and wear that new dress."

Mr. Parris shut the door and walked down the hall. Hanna laid on the bed, still looking out the window, listening to his footsteps. Hanna started to cry, because she was so scared to play tennis with him. She knew from past experiences that he would try to touch her or take it even further. Hanna's mom had brought home so many men, and she just could not handle what she thought would happen after dinner. The tears kept rolling down her face, and as she lay there, she started playing with a little lock on the window, sliding it back and forth and back and forth again, and suddenly realized the window was unlocked. Hanna began to pray. The concept was so foreign to her, but she prayed the best way she knew how, and asked God to help her find a way out of this house and to a phone to call Abi.

Trisha had prayed a few times in an attempt to stop drinking. She told Hanna that Jesus had forgiven her sins, yet she continued to drink and do drugs. Recently, Hanna saw a man outside the Department of Child Safety office praying, sleeping on the ground. He told her he would pray for her sisters and her as they got into two different vehicles the night the Department of Child Safety took them from their motel. Hanna did not know much about praying except that it was supposed to help with everything.

"Please, God, help me to leave this house and find my sisters."

Hanna glanced at the bedroom door to ensure it was shut, then reached for the window and slid it open. No screen was on it, so she knew she could slip through the window. There was a barn down the road that belonged to the Parris's, and she rarely saw them down there. Hanna had a plan. She figured she would hide there until the hired help came to feed and shoe the horses. She would then tell one of them to help her and have them call her caseworker.

Hanna jumped out of the window and ran as fast as she could towards the barn; she did not look back. She was cautious as she approached the barn. She found an open window at the back of the barn, but it was high off the ground. She looked around and did not see anyone, so she grabbed an old white bucket to stand on to allow herself to crawl through the window. Hanna stood on it, crawled through, and jumped down into the barn with both feet. Nobody was around, and it was very quiet, except for some grunting from a pot-belly pig and some horses standing in the stall at the back of the barn. She took a second and thought to herself how nice it would have been to live on a farm with farm animals and her sisters. "My sisters would have loved it," Hanna thought.

"Hi, um, you don't look like another horse to me, and you definitely aren't a pig!" laughed a boy who appeared out of nowhere. This boy's name was Sal, and he was twelve years old. He was a very strong, nice-looking kid, and Hanna could tell he was a hard worker, because he was dirty and had muscles in his arms. She had not seen muscles much, so to her, his arms looked amazing.

Sal was a young kickback boy who lived down the road that the Parris's had hired to help them around the ranch. His family lived down the road and was very prestigious. They believed in the hard work ethic and hoped it would rub off on Sal. Sal's father was not only a rancher but an attorney, and his mother was a pediatrician.

Mr. Parris thought Sal got distracted easily, but loved his hard work ethic and his parents' money. When anything happened, Sal's parents were the first to offer their help, which included their pocketbook. Mr. Parris decided to keep Sal as one of his farmhands and teach him many of the ropes along the way so that Sal could one day run the entire ranch with his parents by his side, offering their help.

"Nobody knows I am here!" said Hanna in a panic.

Sal looked confused, but said, "I am just trying to keep my job, I don't see anything. What is your name?"

Hanna put her head down and whispered, "Hanna."

"I am Sal, Hanna. Are you staying with the Parris's, up at the house? I know they take care of problem kids."

"Yeah, I am a problem foster kid, and I live with them," Hanna said sarcastically.

"Oh, I don't know what a foster child is," said Sal.

"It is when you get taken away from your mom," said Hanna, looking worried and impatient.

"Wow," said Sal, "taken away? Wow, that is huge and different. Why? Who took you?"

"The cops," said Hanna.

"Why?" said Sal.

"Because my mom got in trouble, and I really don't want to talk about it."

"I am twelve. How old are you?"

"I am eleven, I just turned eleven."

Sal looked a little confused but very concerned. "You are eleven years old and living there by yourself with them?"

"Yes," said Hanna, "I don't think they are very nice." Hanna did not mention her sisters that afternoon. She did not know this boy and did not trust anyone as it was.

Sal started throwing hay over to the horses. He did not quite know how to talk to this girl who was a "foster child." Sal leaned the pitchfork up against the side of the barn. "Is something wrong up there at the house?"

Hanna looked down and said with a nervous voice, "I am really scared of them."

"I don't know them that well, I just work here, and the head rancher gives me a list of chores and pays me cash," said Sal, trying to befriend Hanna.

"It's Mrs. Parris. There is something wrong with her. She is mean to me, and I don't trust Mr. Parris. He keeps coming into my room and watching me sleep."

"How do you know," said Sal, "you are asleep," he chuckled.

"Stop! Don't!" Hanna was upset and stressed out, and she knew she had very little time to get out of there before they would come looking for her.

"Hey, I am sorry, what is your name? You said Hanna?"

Hanna nodded.

"Let me try to help you, Hanna. What is it you need me to do?"

"I need my caseworker to know what is going on and how cruel Mrs. Parris is, and how Mr. Parris seems like he is watching me."

"It's ok, really, I will talk to my parents and get your caseworker over here."

Hanna felt a little bit of hope, but still could not relax.

All of a sudden, the barn door opened. It was the head rancher, Lindman. "What is going on here?"

"Hey, Lindman," said Sal. "I was just talking to Hanna."

"Hey there, little lady," said Lindman, as he looked her up and down slowly with a slight grin. "Sal, you need to finish up here and I have some cash for ya'll, and you need to get on home with your family. Have yourself a great weekend and shut that barn door as you leave."

Sal stumbled with his words a bit and said, "Um, that is ok, I will stay and work a while." Sal was deeply worried about Hanna but didn't want to say anything to Lindman. He knew how pushy and condescending Lindman could be, and he didn't want to escalate the situation.

"Sal, it is Friday, here is your cash, I almost forgot, we will see ya'll on Monday bright and early," said Lindman. "Thanks for your help, son; always make sure you do things how I tell ya'll. I have been doing this for a long time, and I know I have bred more cows and bulls than any rancher you will ever meet."

Lindman handed Sal the cash. As he took it, Sal glanced at Hanna, giving her a worried look. Hanna motioned Sal to go ahead and leave, knowing that Lindman was presumptuous and wanted Sal out of the picture so he could brag about himself.

"Hey, Hanna, do you want to see the path I created outside the barn? It leads to this tree house I built, and it is so cool; come on!" Sal gave a tug on Hanna's shirt for her to follow him out the barn door.

"Sal! I said go!"

Hanna was scared and did not move. Sal looked perplexed and hesitantly walked out the barn door with his head down. He was worried about Hanna, but he was only twelve, so he decided to get his parents involved.

"So what is your name, dear?" said Lindman.

"Hanna."

"What are you doing here in this barn?"

"I live with the Parris's."

"Oh, are you one of their foster kids?"

"There is only me right now," said Hanna.

"Do they know you are up here?"

"No, I need to get away from here. I am scared."

Lindman looked at Hanna up and down slowly, as he said, "A little gal your age should not be running away from your home."

"It is not my home. I need to call my caseworker."

Lindman walked closer to Hanna and said, "Hang on, let me see if I have my phone."

He checked his front pocket, and as he was checking it, he unzipped his pants.

"You know, little one, if you will do me a favor, I might help you."

Hanna began to shake, and tears rolled down her cheeks. She knew what was about to happen and started to run to the barn door, but Lindman caught her and pushed her down to the floor, getting on top of her. "You tell anyone this happened, and I'll make sure you never leave this home. I will continue to hurt you."

Lindman put his hand over Hanna's mouth and pulled down her panties with his other hand and lifted himself on top of her, and continued to rape her. Hanna struggled to get away, but he pushed her head down by putting his lips over hers and shoving her head back to the floor. After it was over, Lindman got off her and walked over and picked up her panties, and brought them over to her. Hanna was still crying, shaking, and was just devastated. She thought it was a nightmare that she could not wake up from.

Lindman zipped up his pants, grinning at Hanna. "You were a fresh start to my day, little lady. I can't wait for the next time," he muttered.

Hanna turned away, crying with her hands over her face, wondering how she was going to tell someone what happened without him hurting her again.

The barn door slowly started to open. I quickly put on my panties and prayed it was Sal. The barn door was a bit stuck, and somebody was on the other side trying to push it open. I then saw the end of a rake or broom shoved under the door to help open it. "Sal?!" I yelled, "Help me!"

Lindman looked at me with a threatening look, but I did not care. I had to get some help and let someone know what had happened. Lindman whispered, "Remember what I told you, Hanna."

I could not handle anything more, so I closed my eyes and prayed. "Dear Jesus, please help me."

Lindman stood in silence, watching as the door opened.

Sal's family lived about two miles up the road from the Parris's. They lived in a two-story home in the middle of the beautiful country pines. Sal's parents were professionals, and Sal was raised by nannies. Sal was an only child, and even though both parents worked many hours, he was loved and had everything in life that most boys his age would dream of. He was very close to both of his parents. When he left Hanna in the barn that afternoon, he already knew he was going to tell his parents what happened and find out if they could help her.

It was five-thirty in the evening, and Sal's mom, Sydney, pulled into the driveway. She had just gotten home from work and was very tired after a long day at the office. Sal ran outside and over to her car, and as she opened the door, he impetuously yelled out, "Mom! I have to tell you something! There is a girl named Hanna who lives at the Parris Ranch. She is a foster child, and she told me what that is! Mom, she needs to call her caseworker!"

"Sal, you need to calm down! What are you talking about?"

"Mom! She says that the Parris's are scaring her! We need to help Hanna!"

Sydney looked at Sal with exhaustion, but tried to understand without being too impatient. "You know, when I get home, you don't hit me up with everything at once! What are you talking about, son? Slow down."

"I know you are upset, but the Parris's are good people. They are trying to help foster children. They have a beautiful ranch, and they hired you for Pete's sake."

"No, Mom! They did not hire me; the head rancher hired me. They do not even know me; I have never met them, not even once!"

"Ok, ok," said Sydney, "we will talk to your Dad, he will be home by six."

Sal ran back into the house as his mom grabbed her purse and the few bags in the trunk of the car. She walked into the house and set her purse down on the kitchen counter along with the bags. Sal stood by the back door looking anxious.

"Sal, I know you are upset, but we don't want to jump the gun here. You don't know this girl, and it is nice of you to want to help her, but this can upset a lot of people."

"Mom! The girl jumped out of her window and ran to the barn where I was working. She needed help! I saw her and talked to her! There is something wrong with her!"

"Well, I don't want to call the police or anything just yet, Sal, because I don't know what is actually going on. Let's wait for your dad. How about I call the Parris's?"

"No!" said Sal, "they can't know she is in the barn. She is scared to death!"

"Sal, I don't know how to help here. I don't even know how to approach this."

"Mom! Call Dad!"

"Sal! He will be home in a few minutes. Now, calm down!"

Sal ran to his bedroom and slammed the door. He immediately started pacing back and forth, trying to figure out how he could go back to the barn and talk to Lindman about what was happening. Maybe he could help him.

Sal's Dad pulled up in the driveway, and Sidney caught him as he opened the front door. "Riley! Go talk to your son! There is something horribly wrong and bothering him!"

Riley looked worried but did not ask any questions as he trusted his wife's judgment on everything. Riley was a very tall and muscular man. He was very successful and sure of himself. He did not put up with much. He adored his family and believed right was right and wrong was wrong.

Riley walked briskly down the hallway as Mama had spoken, and he needed to find out what was up. He opened Sal's door. "Hey buddy, what is going on? Your mom is about to burst at the seams."

"Dad! You have to help me."

Sal explained that Hanna had run from the Parris's house by jumping out of the window, and that she was scared and wanted to call her caseworker.

"Caseworker?"

"Yes, Dad, she is a foster girl. She says the Parris's are scaring her. I did not talk to her for long, but I could tell she was afraid, Dad. I think there is abuse of some sort."

Sidney walked into the room. "Riley, we have to help her. We really don't know the Parris's that well. Maybe there is a problem."

Riley, Sidney, and Sal all had a very tight family bond, and each of them would do anything for their friends or each other. Because of this, Sal's parents knew that Sal would not ever be able to let this go without some type of intervention.

"Ok," said Riley, "let me go over to the barn and assess the situation and see if anything needs to be done."

"Thank you, Dad! Thank you so much!"

Sal was so happy and felt that he was helping his new friend without getting involved himself. He feared the Parris's would think he was sticking his nose in their business when he was just a hired hand. He felt relieved and hopeful.

"I will be back, Sidney. Don't hold dinner up for me."

Riley opened the back door and got in his car, and drove over to the Parris's ranch. The barn was located up the dirt road from their house, and there was no fence, just a lot of trees blocking the barn. The house was on ten acres of land, but set back from the main road about a mile. It was secluded behind a lot of trees, and most would not even know the house existed, because it was hard to see from the main road.

Riley's truck was new and did not make a lot of noise as he approached the barn. He decided to enter through the back barn door. He saw Lindman's parked truck. It was gray and dented with chipped paint. Riley got out of the truck and walked towards the barn door. He tried to push the old barn door open, but it was stuck. He saw a broom with a wooden handle on the ground, and he picked up the broom and used the handle to help pry open the door. The barn door squeaked, and he slowly walked into the barn with apprehension, as he believed something was going on, but was not sure what.

"Hello? Lindman? I need to talk to you about Sal, man. Where are you guys?"

All of a sudden, Riley heard a radio playing in the back of the barn. He walked towards the back, and when he turned the corner, he tripped over a wheelbarrow that was thrown to the side. Lindman walked out of the back of the barn. "Hey Riley, how are ya'all doing?" said Lindman.

CHAPTER: 7

Safe Havens and Silent Cries

"Sami! Straighten your right leg. There you go, hun," Emma, who was also known as Mrs. Black and Sami and Lindsey's foster mom, was trying to secretly instruct Sami without the coach knowing. She was commonly known for stepping in and trying to take over, using the expertise that she thought she had. Sami was trying to somersault and was having more fun tumbling over and over, as she let out giggles.

"Sami, you can do this! Come on, tuck your head!" said Emma. "Just roll with it!"

"Emma, Sami might not be able to tumble straight over because of her feeding tube," said Lindsey.

Lindsey really liked Emma and felt safe with her, knowing she was in good hands. "Oh, she is capable of anything. That child is brilliant!" laughed Emma. "Now you get out there and show me your stuff, little lady!" Emma laughed as she nugged Lindsey.

"No, thank you," Lindsey smiled. She was happy just watching Sami for now. She knew Sami was getting stronger by participating in gymnastics. Lindsey was worried about Sami, and she tried not to show it. Sami had not gained a pound since they arrived at Emma's house.

"Lindsey, I am waiting to hear from Friends of Foster Care," said Emma. "I want to see if they will pay for Sami's gymnastics. It's not that I can't, honey, but that is what the help is for, to help you kids get what you need. Foster parents need help to make your visit with us a great experience, but it is always nice to get additional help." Lindsey nodded her head in approval. Emma was used to doing five million things, multitasking, and wanted to put both girls in everything. She was enjoying it more than they were!

Emma felt guilty for not putting Lindsey in gymnastics with Sami. She and Canby had planned on paying for it one way or another, but wanted to see if it was possible to get some assistance through Friends of Foster Care first.

Emma and Canby made a great income, and they definitely were not doing foster care for the money. They were concerned about the girls and very upset about the fact that Hanna was living somewhere else. They were asking questions to other Foster Parents about their experience with siblings. They seemed to be hitting a brick wall because Hanna was already placed, and Trisha, their mom, was trying to fight to get the girls back.

The Blacks called Hanna's case manager and left two messages, and it had been a week and they had not heard anything from them. Lindsey and Sami knew nothing about their trials to get Hanna, and they did not want to get their hopes up.

"You have cute grandbabies," a woman sitting next to Emma said.

"Thank you, but they're not my grandchildren; they're my foster daughters." Emma was proud of being a foster parent; she was helping the community, and it was so needed in this little town.

"Oh, ok," said the woman. "I could never do it, good for you. My husband would never allow me to do something like this; he works too much, and I need my time to take care of our home."

"Why do you say you could never do it?" Emma asked, her brow furrowing in confusion.

The woman replied, "They have so many behavioral problems, you don't know where they come from. I don't want them affecting my kids, rubbing off on them." Emma looked irritated and puzzled because she could not understand how people could be so close-minded. She decided to kindly ignore the woman and looked towards Sami. The woman walked away.

"Better riddance," thought Emma.

"Where is Sami?" asked Emma.

"Right here." Lindsey was watching Sami and admiring her increased strength and her will to learn. She had never been given any kind of lessons before. "She is over by the drinking fountain, see?"

Just as Lindsey pointed to the drinking fountain, Sami fell to the floor. Both Lindsey and Emma looked at each other. "What was happening here?" Emma looked perplexed. "Is she playing?"

"No!" Lindsey jumped up and ran out onto the gym floor. One of the instructors had beaten her to it and was checking Sami out to see what was wrong with her. The instructor looked at Emma and said, "I don't know what happened; she was fine and just passed out."

"Call an ambulance, my gosh, don't just stand there!" said Emma.

"Wait," the instructor said. "We can call them, but we do have a nurse, you know, the gal with the red hair at the front desk, her name is Linda, someone go get her!"

Linda came running and bent down to check Sami out. "She is breathing a bit rapidly, but no need to do mouth-to-mouth. Her skin is a bit cold. Where is her parent or guardian, may I ask?"

"I am here," said Emma, standing in the back, waving her hand in the air with a frightened look on her face.

There was now a crowd around Sami, and Emma was standing in the back trying to make her way through the crowd.

"Has she eaten today? Do you know?" said Linda.

Emma raised her voice so Linda could hear her. "Well, she is on a feeding tube, and she is a foster child, so I don't know exactly how long she has been on it. The state is always so vague, you know. I do know she does not eat a lot of food ever, but she seems to be very healthy after the filthy life she led."

Linda looked at the feeding tube connected to Sami's stomach and noticed it looked clean and intact. "Well," she said, "she definitely needs to be seen by a doctor. Let the ambulance come and take her for an assessment."

Sami started to wake up as the medics finally arrived and rushed to her side. "Please lie still, honey," one of the medics said.

Sami started to cry, her voice trembling. "Mommy, Mommy, I want to go home." Sami saw Emma through her tears, which made her feel comforted, but she was still scared. The medic tried to calm her down as Emma and Lindsey squeezed their way in through the crowd, so they could be closer to Sami.

"It's alright, Sweetie," said Emma, "we are here with you."

Lindsey put her hand over Sami's and slightly squeezed it. Sami calmed down as the medic explained they were going to take her to the E.R. for some tests.

"Is she going to be ok?" Lindsey asked, as Emma interrupted, "Just let them do their job, Lindsey. She needs to go."

The medics put Sami on the stretcher and walked out of the gym towards the ambulance. Sami and Emma followed him as the instructors returned to their students and reassured them she was ok. Emma ran up to the medic and asked if they could ride with Sami, but then realized she would not have a vehicle to get back home, and her husband was out of town.

"Never mind," she said. "Are you taking her to North General Hospital across town?"

"Yes, ma'am," the medic said. They got in the ambulance and drove off with lights flashing.

CHAPTER: 8

Trisha's Downward Spiral

Trisha was feeling depressed and needed a fix as she sat on a log outside the parking lot of a convenience store. An old blue Toyota truck honked, and some guy in a plaid shirt she recognized as a fellow junky waved as they drove by. Trisha jumped up and kicked the log she had been sitting on as she realized she could have hit them up for some drugs. "Dammit!" she thought. "I have been sitting here all day! I need to feel good! I am so bored!" Trisha started kicking the dirt and slowly walked towards the market and opened the glass front door. She was hungry and thirsty, with no money to her name. She knew what she had to do, so she kept her head down and walked to the back of the store, where the soda refrigerators were. She looked around and saw nobody was watching, so she slid a Pepsi into her purse. She continued down the next aisle and slid a beef jerky and some peanuts into her purse as she smiled at a couple approaching her. Trisha was never a thief, but in this moment, she felt desperate enough to do whatever it took.

"I hope they did not see me," she thought. The couple had walked by her, chatting and laughing as they walked over to get in line at the checkout stand. She did not think they were paying much attention when she took the items.

Trisha kept her head down as she got in line behind them. "Can I help you?" Trisha looked up at the young man standing in front of her, who resembled a store clerk. "No, I am just looking." Trisha put her head back down. The clerk briskly walked by and began stocking items in the next aisle. Trisha hurriedly opened the front door and walked out with her stolen items. "Whew!" she thought. "That was a close call."

There is one thing about Trisha: she always got what she wanted. She was her own priority. Nothing ever mattered at the moment, but she was hungry on this day, and she would do anything to fix that. Trisha started humming a song she heard on the radio that morning. Without a care in the world, she continued down the dirt road, eating her peanuts. There was a run-down trailer on the left that called out her name as she walked towards it. It was her home for now. Trash covered the entire lot, and there was so much junk that it was hard to tell what anything was. Trisha tried the door, and it was locked. She tried it again and began knocking harder. Nobody answered, so she started kicking the door and yelling Rome's name. "Rome, answer the damn door, you fool!" "Come on!" Rome angrily slung open the screen door. "Go away, Trisha, I am busy right now!" Trisha headed up the steps when Rome pushed her back down them. Trisha stumbled and fell back on her left arm. "What the hell?" said Trisha, confused and furious.

Trisha got up and started back up the stairs. "Are you drunk? It's me, you jerk! Move out of the way. I am done with this day, and I want to relax!" "No, Trisha," said Rome. "What? What's wrong with you, Rome? Let me in!" Trisha nudged halfway through the trailer door as Rome tried to stop her by putting his leg in front of her. "Trisha, there is someone here, and you need to leave now." Trisha shoved Rome aside, as she looked to her right and saw a young woman with blonde hair, undressed with a blanket wrapped around her, lighting a cigarette. The woman looked about 18 years old, if that, and was having a hard time standing up. She quickly turned around and went back into the bedroom.

There were only two rooms in the trailer. The kitchen was part of the living area, and the bedroom was in the back, which included a tiny bathroom. Trisha and Rome lived there for free because they were going to clean the place up for the owner, who was a junkie and drunk himself.

"What is going on, Rome?" "Trisha, you need to leave!" "You asshole! This is my house, too!" Trisha quickly jumped from the trailer over the steps, picked up some rocks, and started throwing them at Rome. As Rome dodged the rocks, she continued cursing angrily. Rome quickly shut the door and locked it. Trisha was fuming and was not through yet; she ran back up the steps, kicked the door, and threw a large rock through the trailer's living room window. "Now, see if you can afford the electricity with a hole in your window!"

Trisha chuckled to mask the hurt inside, but the laughter quickly turned to sobs. She sat down on a large rock and just fell apart. After a while, she got her thoughts together and decided Rome was not worth her time anyway. Trisha got up off the rock and headed down the dirt road again. She was upset about Rome's cheating, but also knew that nothing lasts forever, and love is just not in the cards for her. "I deserved it again," she thought. "I just don't have anywhere to go, which is now the current problem. I could go down to Mercy Lane; maybe they could put me up for the night or two." She knew if she did not get to the shelter by nightfall, she would be on the streets without shelter or food.

Mercy Lane was a downtown shelter that took in homeless and abused women and children. Trisha had been in the shelter a few times with her girls. This shelter had goals that the women had to meet, with the final goal being Transition Housing. Trisha never made it to the transition homes. These homes were for women kicking their habits and learning to support themselves and their children. She never qualified because she would sneak out with the kids and go back to her old ways, which were always addiction and prostitution.

Trisha hoped Mercy Lane would take her back in and that she had not burned all her bridges, so she started heading downtown. Trisha started praying as she neared downtown, her thoughts muddled with desperation. God had always been foreign to her, but in times like this, she knew she needed him more than ever. Her grandmother had told her about God, and even though she did not pray much, she knew when she needed him. "God, I've messed up again. I don't know where to go from here, but I need your help." Tears rolled down her face as she knew she would not ask for forgiveness because she was not ready to stop taking drugs. Drugs were the only thing that made her feel good. Trisha arrived at the shelter, and as she walked up the steps, she thanked God for getting her there.

CHAPTER: 9

The Barn of Secrets

Lindman had never been convicted of any crimes; however, he was known for being a womanizer. Riley had entered the barn as Lindman walked out from the back and said, "Whatcha need, Riley?" "I just had some questions about my son's pay." "Yeah?" Lindman said, his tone distrustful. "You know I am pretty good at Math; why are you really here, Riley?" "Ok, my son feels like there is some type of problem with that little girl Hanna." Riley was trying hard not to seem too suspicious, but he really did not have a lot of trust in Lindman in the first place. Riley had caught him in some lies in the past, and Lindman had never apologized or tried to make amends. "I need to see that she is ok, Lindman."

There was a faint sound in the back of the barn like someone was moving around. "Did someone call for me?" Hanna came out from the back of the barn, pointing towards where she was coming from. "I was back there trying to learn where everything goes since I might be working here." "Lindman was showing me some things." In Riley's opinion, Hanna looked a bit frazzled, but he did not know her, so he did not want to jump to conclusions.

"This is Hanna," Riley said. "Hanna, hi, I am Sal's Dad, and Sal was a bit worried about you. I am just gonna cut to the chase here. He thought you seemed a bit scared about living with the Parris's. He thought you were acting funny before he left this afternoon. He had just met you, so he didn't want to jump to conclusions, but he asked me to check on you." Riley tried to lead the conversation, but left it open so Hanna could speak up. "No, sir, I was just nervous because I did not ask to come down here to the barn. I am scared I'm gonna get into a whole lot of trouble." Riley looked at Hanna with genuine concern. He did not believe her, but why would she lie and not tell him what was happening? She did not show any type of concern as of now.

Hanna was used to lying to her mom's boyfriends and customers, so she was good at it. She would fake that she was asleep or sick, so they would leave her alone. Lindman walked over and handed Riley a one-hundred-dollar bill. "Thanks," said Riley. "Sal is working hard on buying a better bike. We can, of course, buy him one, but we're trying to give him some more responsibility, you know." Lindman started laughing and nodded. Riley thought Lindman looked suspicious, but decided to give him the benefit of the doubt, only for now.

Riley looked at Hanna and said, "Well, Hanna, you gave us a scare. I guess I should have thought twice about assuming you would not be safe with old Lindman. I'll tell you what, girl, if you think you are in trouble, you'd better get yourself back up to the house now; better safe than sorry." Hanna nodded with a little smile and looked down at her feet, not wanting to continue the conversation.

"Lindman, do you want to make sure she gets up there, ok? Sal and I would appreciate it."

"I got this," said Lindman. "I gotta talk to the Parris's about their colt anyway. We need to change her food because she is getting colic." "Alright then," said Riley. "Hanna, if you ever need anything, please let us know; we just live right up the road, hun. You can't see the house from here, but it's a big yellow house, two stories up the road on the right. It's the only house down that way. You can't miss it." Riley said, knowing that Hanna might decide to head up that way soon. Lindman looked uneasy as he told Riley to have a good night and to say hello to Sal. He was not happy that Hanna knew where Riley lived.

Riley walked out of the barn and got into his truck, and drove away. Hanna heard the truck drive off as she quickly glanced at Lindman and started walking towards the barn door. She could not wait to get away from him. All she wanted was to run back to the house, take a shower, and escape seeing Mr. and Mrs. Parris. Lindman was supposed to take her back to the Parris house, and he knew it, but he decided to add a bit more fear to Hanna's day. He could not have her talking to anyone about what happened.

Lindman positioned himself in front of Hanna and put his left leg right in front of the door.

"Think you are going somewhere, little lady?" He took his hand and brushed back Hanna's hair. Hanna trembled with fear. She tried to be brave and remembered what her caseworker told her when they arrived at the Department of Child Safety Office. "I am the only one who can change my future; it is my decision how I want to grow up." As Hanna thought back, she became angry and pushed Lindman's hand off her head and plowed through the door with all her strength, as fast as she could. Lindman just let her go. He could see she was fighting back and decided he had better lie low.

Hanna ran down the dirt road back to the Parris's house as fast as she could. She ran so fast that she fell and rolled into a small muddy ditch. She picked herself up, brushed her pants and shirt off, and continued running towards the house. She wanted to sneak in, but she had been gone most of the day, and she knew she would be in trouble. She thought that if she made up a story about the barn, she might get away with it.

There was no way she was going to mention the rape, because they were weird people themselves, and she never knew what to expect from either one of them. She definitely did not want to make things worse. She decided to wait for her caseworker, tell her everything about Lindman and hoped he would be arrested for what he did to her. Hanna could not bear the thought of him hurting another child.

Hanna was back at the Parris house. She slipped silently through the back door that led to the kitchen. She was nervous, and as she closed the door, it accidentally slammed behind her. "What are you doing?" It was Mrs. Parris now standing right in front of her with an annoyed look. Hanna was out of breath from running so far and being so stressed out. "I was outside enjoying the day, ma'am, and I lost track of time while looking at the barn." "You are such a lying disgrace! You were not outside enjoying the day, child! Good Lord, you were gone most of the day and have nothing to say for yourself but a downright lie! Get out of here, child! Go wash up for dinner!" "I thought you would want me to shower real quick before dinner since I had been out all day," said Hanna. Mrs. Parris walked into Hanna's bedroom and stood between the bathroom and bedroom door, wanting to continue yelling. "Good Lord, all right, child," said Mrs. Parris. "Just come to dinner, for God's sake! You are really turning into a problem, and we need to talk about this at dinner." "I will hurry ma'am, thank you," Hanna said quietly, as she knew she was going to go through more hell that evening.

Hanna ran down the hall into her room, darted into the bathroom, and quickly shut the door. She grabbed some toilet paper, wet it, unzipped her pants, and wiped off her privates with tears rolling down her face. "I just want to take a shower," she thought. She quickly threw the toilet paper in the toilet and flushed it. "If I do it fast, nobody will notice." She threw her clothes off and jumped in the shower as fast as she could.

She soaped her body down and cleaned her privates, still crying, shaking, and thinking about what had happened that afternoon in the barn. "I just need to talk to my caseworker. I need to get to my caseworker; she will help me." Hanna heard the bathroom door open. "Please don't let it be Mr. Parris." "Hanna, what in the world are you doing now?" It was Mrs. Parris, and this time she sounded very angry.

Hanna finished her shower and got dressed with the little choice she had in clothing. Mrs. Parris had put a few things in her drawer when she had arrived. As she walked down the hall, she passed Mr. Parris' office. The TV was on, and the chair he was sitting in was facing the door, so she quickly walked by, hoping he did not see her, as he was immersed in the football game on his big screen. "Hanna, do you like football?"

Hanna cringed at the sound of him, and she did not back up to the doorway where she could be seen. "No, sir," Hanna was apprehensive. "Come on in and watch it with me for a while, hun. I have been here for a few hours and would love some company. You might start to like it." Hanna stayed still and did not show herself. "No thank you, Mr. Parris. Dinner is about to be ready, and I am really hungry."

Mr. Parris laughed and got out of his chair and walked to the doorway. He turned to Hanna. "Hanna, get in here and watch the game with me for a while, shut the door so we can hear." "Dinner is not ready yet, believe me, she will holler when it is, we know that!" Mr. Parris laughed. Hanna slowly backed up to the doorway where Mr. Parris stood staring at her. "Sir, I really don't want to come in here and watch football." Hanna did not trust Mr. Parris one bit, and she knew he would try to touch her or do something to her if given the chance, and this seemed like the chance.

Hanna was scared, but walked in front of Mr. Parris through the doorway and into his office. Feeling like she wanted to throw up, she faced the TV and did not move. "Hanna, come sit by me. There is a lot of room on this leather couch, hun. You don't have to stand up and watch the game." Hanna turned around and looked at Mr. Parris, trying not to show fear in her eyes. She was stronger than this. She could come up with an excuse and get out of this. "This will not happen," she thought. She wanted to break down and go to pieces because of what Lindman did to her a few hours ago, but she held it together.

Mr. Parris motioned Hanna to come over and sit by him. He patted the leather couch with his hand and smiled. "You will like this game, hun. Come on over." "Mr. Parris, I feel like I pulled a muscle today, and I need to talk to my caseworker about it." Hanna knew what she was doing. If Mr. Parris tried to touch or hurt her, she wanted him to know she was already going to talk to her caseworker. "That's okay, hun, I can help you with that. Come on over here and lie down on the couch, and I will rub whatever hurts for you. You know, massage it out for you."

Hanna was petrified and decided she would run out of the room as soon as she could and scream, hoping Mrs. Parris would take her side, but the office door opened. "What is this door closed for? Can someone explain this to me?" It was Mrs. Parris, and she did not seem happy.

"You know this girl is a foster child, and the rumors that can start! What are you thinking?" Mr. Parris sat in silence as he knew he was caught red-handed. He also knew he could talk his wife out of just about anything. "Hanna, what are you up to, child?" said Mrs. Parris. "Go to the table now! Dinner is ready!"

Hanna ran out of the office to the kitchen and sat down at the table. She knew she would get in trouble for not washing her hands, so she walked over to the sink and washed her trembling hands. She walked back over to the kitchen table and sat down in fear of what she thought was about to occur when Mrs. Parris came back. She decided to pray. She prayed for the strength to be able to handle what she thought was about to happen. "God, please give me the strength to be able to get through all of this. It seems like it is a horrible nightmare, and it is really happening. Please God help me."

Back in the office, Mrs. Parris was pacing back and forth. "Ok, do you want to explain yourself?" asked Mrs. Parris.

Mr. Parris got up off the couch and stood before Mrs. Parris like he was being judged by a jury. "Martha, she comes in here and bends over in front of me, trying to see my reaction. I am getting tired of it. I tried to ignore her. Christ, she shut the door this time and tried to come over and lie next to me. I was trying to rest and watch the game! I told her to sit in the office chair if she wanted to watch the game with me while we were waiting for dinner. She bent right over in front of me with her bum right in my face. She was trying to get a reaction out of me. It is like she has been trained as a stripper. What is wrong with her? You don't do that to a man. A happily married man. I want her out of here, damn foster child. I hope you believe me, babe."

"I am going to call her caseworker and have her removed," said Mrs. Parris. "I don't care for this one at all! She is out all day long and comes in filthy dirty saying she was out enjoying the day, and then comes in here and decides to enjoy my husband! You can't trust these kinds of children! I will teach this little piece of crap a lesson!" "Be careful," said Mr. Parris. "You show a mark where you spank her, and we could get arrested. You know how the system is. You can't spank a child." "I know, I know. I know exactly how to take care of this foster brat!"

Mrs. Parris had an idea that would hurt Hanna very badly and make her learn. She did not feel guilty or bad about her thoughts because she knew Hanna was trying to move in on her husband. She deserves whatever she gets, she thought.

Mrs. Parris left the office, premeditating her plan in anger. She was cooking fried chicken in the kitchen. She had put a lid on the chicken to keep the grease from popping while she checked where everyone was and why they were not coming to dinner. "Get up and help me, child!" Hanna got up feeling very queasy and dizzy; not only was she exhausted and had not eaten, but she was scared. "Yes, ma'am." "Help me with this fried chicken. Here, take the tongs and turn that chicken over." Mrs. Parris was being very persistent with her request.

Hanna backed up because the grease was flying all over, and it stung her arms. "Mrs. Parris, can we turn the fire down?" Mrs. Parris pushed Hanna towards the stove. "Get in here and learn to cook! You can't do much else but get yourself into trouble, so learn!" Hanna had the tongs in her hand and started crying as she turned the chicken over. The grease was splashing on her hands and arms. Mrs. Parris started yelling, "That's it, turn it over, turn it over!" "It is hurting me, Mrs. Parris!" Hanna backed up from the stove and threw the tongs in the skillet.

Mrs. Parris grabbed the hot skillet with a pot holder, picked it up, and slammed it back down again. "I said, turn the chicken over, foster child!" Hanna was sobbing, but picked up the tongs as Mrs. Parris grabbed the skillet and dumped the hot grease all over Hanna's right arm. Hanna screamed and grabbed her arm, sobbing as she fell to the ground. "Oh dear Hanna, I didn't mean for this to happen! I was just trying to teach you to cook!"

Mrs. Parris grabbed Hanna by both arms, not caring if she touched her bad arm or not. Hanna was barely able to stand up, but Mrs. Parris held her up as she attempted to turn the cold water on and rotate Hanna's arms under the water. Hanna passed out.

CHAPTER: 10

Scars of Silence

It was early morning, and the sun shone through Hanna's bedroom curtains. She awoke to horrible stinging and pain. She could not move her right arm at all. It was bandaged up. She did not remember how she got to her bedroom or who put her in bed. The bandage ran from her wrist up to her shoulder. The pain was so severe that she dreaded asking for pain medicine, yet she needed something. She did not want to ask Mr. and Mrs. Parris for anything. Hanna remembered how it happened, and she knew Mrs. Parris did it on purpose. God only knew what they would give her if she asked for an aspirin.

The door opened slowly. "How are you, dear?" Mrs. Parris asked. Hanna looked up at her with a feeling of revulsion. "I want to call my caseworker. I need to talk to her now," Hanna said.

"Why?" Mrs. Parris replied. Hanna sat in a distrusting silence.

"Well, my dear, we must discuss a few things first. I need to make sure you understand what happened to you. I am going to let you rest today. I will fix you some eggs, bacon, and toast.

You did not eat last night. You were so tired, and although it made no sense due to your exhaustion, I went ahead and let you help me fry the chicken. Now, look what happened. Do you remember, child?" Mrs. Parris was doing her best to manipulate Hanna. "You were exhausted and kept leaning on the stove, too close to the burner. I kept telling you to back up because you would get burned. Do you remember? You picked up the frying pan and tried to move it to another burner; somehow, the grease spilled all over your arm. It all happened so fast. You then fell to the floor, and I tried to help you. I picked you up the best I could; you are not light, you know."

"I put both of your arms under cold water because I did not know where the grease had hit you. You then passed out. Mr. Parris and I moved you to the bed; you were just so exhausted. We saw the burn on your right arm, applied some ointment, and wrapped it up to make it as good as new for you. We care about you, Hanna. We made sure you were safe in bed." Hanna did not say a word. "Now I am going to fix that breakfast; you rest." Mrs. Parris walked out of the bedroom and down the hall as Hanna lay in bed, trying to figure out what to do and how to get to the phone without Mr. and Mrs. Parris finding out.

Hanna was so sore, and her arm hurt so bad, but she got up and got her paper dolls. She stood a few against her bedroom wall while she lay in bed. She started talking to a few of them. "It is not fair, but we have to be strong. I can't let this get to me. We have to get to the girls. I don't know where they are, and I have to find out. Help me be strong." Hanna bowed her head and asked God to please help her. She put the paper dolls back in the bag and slipped them back into the drawer so Mrs. Parris would not see them.

CHAPTER: 11

Prayers in the Waiting Room

Flu bugs filled the emergency room, as the local druggies complained that they needed meds for anything they could think of. Lindsey and Emma walked in and looked for an empty chair at the registration desk. Lindsey found one, and they both sat down. Emma saw that one of the clerks was on her way back to her seat. "Is this where we start the insurance information process? I think you may have our information in your system already. Can you check, please?"

The woman behind the desk asked for Sami's insurance card and information. "Thank you, you may have a seat in the waiting room." Emma and Lindsey walked towards the waiting room, found a seat, and sat down. "Good Lord, Lindsey, they had to have beaten us here. I wonder if they put her in the intensive care unit?" Emma got up and went back to the information desk. "Excuse me, ma'am, but can you tell us if Sami Flemming has arrived by ambulance yet, and where she has gone?" "Ma'am, please have a seat, and someone will be out in a few minutes to give you the information you are waiting for." Emma turned around and angrily stomped back to Lindsey, plopping down in her chair with her arms crossed.

"I can't believe this happened to her," said Emma. "She loves gymnastics, and I just don't get it. Would she have episodes at your house?" Emma asked. Emma did not get much information from the Department of Social Services, but she did know that the girls had lived in a motel among other places, and there was no stable home environment. "I don't really remember, Emma," Lindsey stuttered, "I mean, Mama, I don't remember because Hanna always took care of us." Lindsey had tears in her eyes, and Emma knew the absence of Hanna, and now the episode with Sami, was hard, especially when Hanna was mentioned.

Emma decided to be positive and hopeful for the girls. "Well, honey, I do know Sami is in great hands. This is a reputable hospital, and she's in good care." Lindsey was so young, but there were so many adult topics that needed to be discussed, and Emma just kept trying to communicate and find out what she could from the girls in her daily conversations. "I also know that your sister Hanna is in God's hands and she is being taken care of while the judge figures all of this mess out." Little did Emma know that Hanna was struggling and being abused where she was being fostered.

Emma was sad the girls were separated, but she was going to protect Lindsey and Sami above all, because they had not had the physical and emotional abuse that Hanna had. Hanna took all the abuse sexually and mentally to protect them. "You don't want to be around Hanna right now, as she went through a lot while living with your family. She needs to get some help before there is any reunification with you girls, that is for sure." Lindsey did not quite know what she meant by "reunification," but she knew she was happy living with Emma, and she trusted her so far.

"Mama, I can't lose her. Without Hanna and Sami I have nobody," said Lindsey. Lindsey had tears slowly running down her face. She kept wiping them away. "I don't know what happened to my mom. I miss all four of us together. I love us." Emma wondered what Lindsey meant. "I loved all four of us together." Did she mean their mom? How could she love that mother after what she put her girls through? She allowed her oldest daughter to be sold for drugs and raise her younger siblings. How could a child love someone like that?

Emma put her arms around Lindsey. "My sweet, beautiful child, you are not alone. You have me, and you have God. You must pray when your burden is too much for you to handle and ask God to help you. Like this, honey…" Emma took Lindsey's hands in hers and closed her eyes as Lindsey watched with tears continuing to slowly fall. "Dear Jesus and precious Father, our sweet girls are in need of your help. These girls, Lord, have been through so much. Please be with Lindsey, Sami, and Hanna and help them overcome whatever pain may be with them as of this time. Please give them the strength they need, dear Lord. They need you and they want to know you. Please give Lindsey peace as she awaits her sister's diagnosis from the doctor and peace that her sister Hanna is safe and healthy. We know, dear Lord, you are the great physician and can heal Sami. Please, dear Lord, be by Sami's side as you heal her and give her your love. We love you, dear Lord, in Jesus' name we pray."

Emma looked down into Lindsey's eyes and with all the love in her heart, she said, "And that is how you pray, my love." Lindsey was amazed at what she just witnessed, but she knew that Emma meant every word she said, and Emma's beliefs and prayer felt good to her. Lindsey decided she wanted to find out more about this God.

"Emma Black?" said Dr. Checkers. Emma stood up and nodded her head in agreement. "Sami is dehydrated, and she needs to drink more liquids." "I will make sure she does, doctor." "Yeah, it's hard because of her situation," said Dr. Checkers. "That feeding tube absorbs a lot of liquid. She has to drink and eat twice as much to gain weight and keep from dehydrating. Sports in themselves make a child extra thirsty and hungry."

"I understand, doctor. When can she go home?" "You can take her in a few hours tonight." "Really?" said Emma. "Yes, just let her rest a bit and go get yourself something to eat upstairs and come back about 7:00 PM, she should be ready then." "Dr. Checkers, is that all that's wrong? Just dehydration?" "That's all I can see for now," the doctor replied.

"You know it is great that she does gymnastics, but she does wear a feeding tube, and it needs to be watched and handled with care." "I understand and agree with you, doctor. We clean it properly. Is there anything else we can do to help her at this time?" "I would have her rest every afternoon for an hour or so and drink a lot of fluids. She does not need to stop gymnastics, but her coaches need to be aware of that tube, and they need to make sure she is hydrated as well." Dr. Checkers gave them a nod and turned and walked away.

"Well, thank God she is ok," said Emma. She smiled at Lindsey as Lindsey hugged Emma. "Let's go upstairs, hun,

and get some dinner and go pick up that girl of ours!"
Emma and Lindsey left the waiting room and walked down
the hall to the elevator. They were both hungry but relieved
that Sami was all right.

CHAPTER: 12

Mercy Lane Revisited

Mercy Lane was a shelter known for helping abused and drug-addicted women and their children. Trisha had been there quite a few times in the past with her girls, but was kicked out because she did not care about helping herself. She went there to socialize and brag about her drug use, never involving the girls in the supportive activities available for them. The administration used the reason that there was a time limit to stay at the shelter, and her time was up. Trisha was usually fed up with the shelter and needed a fix so badly that she was happy to leave.

She would just take the girls out of the shelter and walk down the street with them no matter what the weather was, no matter what time it was. Trisha did not care. She would flag down vehicles and she and the girls would go to unknown places for the night. She would always figure out a place to land until she could talk someone into getting her another motel room. She always seemed to escape The Department of Child Safety because she never stayed at the same place long.

The director of Mercy Lane was Angie Long. She was exceptionally dedicated to the women and children of the shelter. She was known for being very strict with the rules and not tolerating much. It was cold and dark now, and

Trisha knew she had to find a way to stay at the shelter or she would end up being alone in the cold, sleeping on the streets somewhere. Trisha always depended on her men to support her since she had no job skills and could not stay sober and off drugs. Sex was her payment for everything, and when that was not enough, she offered Hanna at higher price.

Trisha knew it was not safe to walk around town at night, especially in this neighborhood, even though she spent many nights strung out walking around, but being sober was a different story. She actually feared someone would hurt her. Trisha opened the front door to the shelter and immediately saw Angie Long standing in the kitchen. Angie pretended not to see her. It was not that Angie disliked Trisha; she was just tired of taking her in and having her not take her admission seriously.

She would continually look for someone who would have drugs or put her in contact with someone who did. Angie hated seeing her hurt her children by taking them in and out of the security the shelter offered. Trisha did not want the chance that she was offered at Mercy Lane. She knew she could get transition housing for her and the girls, and it did not even matter to her. It was too hard to quit her habits. She would rather deal with the consequences than quit; it was easier.

Angie called the Department of Child Safety a few times because she thought the kids would be better off. By the time they would arrive Trisha would always be gone. It is like she knew something was about to happen, and she had to get out of there.

Trisha walked through the living room and passed by numerous women and children. Feeling sorry for them was not in the picture as of this time, since she had not had a drink or any drugs for over seventy-two hours. Trisha was nauseated and felt horrible and wanted to find a place to sleep and see if anyone at the shelter had a fix for her. There was always someone living at the shelter who could hook her up.

Trisha walked into the kitchen as Angie looked up. "Now what, Trisha?" said Angie, rolling her eyes. "Hey, Angie, you got room for me?" Trisha started jabbering off at the mouth. "I have been doing so good, but my boyfriend, you all know, he found some other bitch and locked me out of my own house, Angie, and I have nobody, nowhere to go, man." Trisha was crying. Angie was already fed up. "Nope," said Angie, "I don't have room for you. You need to leave, Trisha, and you know why." "I know," Trisha said, "but this time I am trying really hard. You burned your bridges, and the answer is no! And there is the door!" Angie pointed to the door as Trisha stopped her feet walking towards the door. "Angie, please. It's getting dark and I am scared in this neighborhood. I promise I will be gone by morning." "Trisha, I can't give up a bed when others really want recovery."

"If you don't leave now, I am going to have to call the police, okay," said Angie. "You did not even ask about my girls." "Your girls are not with you, Trisha, and I have heard that they are in foster care!" "How do you know, Angie?" "Trisha, I don't believe in turning anyone away, but you don't care about you or your girls! You don't want to help yourself, and it has been too many times that you come in here and beg for drugs and money from the other mothers.

Your kids are always filthy and hungry, and I get that you need help, but you don't want to try at all to better your life. I can't help you if you do not want help!" Trisha held up her middle finger, flipping Angie off, and turned towards the door and slowly started to leave. As she looked back, several women and children were watching her.

"You will kick all of these people out too, it's just a matter of time, Angie. It is just a matter of time!" Trisha yelled to everyone.

Trisha pointed to herself in anger. "I do better than all of you do! I am better than you scum bags!" Trisha was crying as she shut the door behind her and walked down the steps and back outside the shelter. The only other place she knew there would be others like her and possibly some drugs was down by the bridge, and everyone called it Drug City. It was considered very dangerous, but all she needed was a fix, and she could be on her way to find somewhere to stay for the night. There was no way she was staying under that bridge. Trisha started walking briskly towards the bridge; it would be dark soon.

CHAPTER: 13

Escape from the Parris House

"Why is there a burn on her arm? I need an answer," said Mr. Glasso. Mr. Glasso was still assigned to Hanna's case, but only for a few more months because he was transferring to another county, which Hanna did not know about. Hanna had waited until the Parris's went to sleep after having the hot grease poured on her arm. She had stashed away her caseworker's card in one of her socks in her drawer. As much pain as she was in, she got up and snuck into the office where the home phone was connected. She called the Department of Child Safety Office and told them it was an emergency and that her foster mother had poured grease on her arm. Mr. Glasso just happened to be in the office doing paperwork. He left immediately for the Parris house with the police not far behind him.

Hanna had to be careful because she wanted to tell Mr. Glasso about the rape in the barn, but she thought it would be so overwhelming to bring it up at this time, so she decided to wait. She thought Mr. Glasso would have a hard time believing two absolutely horrible things happened to her back-to-back. She decided to wait until she found out what would happen to her after this incident.

Mr. Glasso was very upset and adamant that he receive an answer immediately. I asked you why there was a burn on her arm. I told you, sir, I tripped over the rug in front of the stove, and my arm hit the frying pan, and the grease splashed up on Hanna's arm. I've already told you this twice! Mrs. Parris was so good at lying that she actually seemed trustworthy. "Hanna says that you were angry with her and you threw it at her! Is that what happened to Mrs. Parris?" Hanna put her head down, not wanting to participate at this moment. Okay, Mrs. Parris, I have no proof, I will give you the benefit of the doubt, but I need to find out what was going on over here for you to throw grease at this child! It was not thrown! I tripped! Don't you dare blame me!

"Hanna come here," said Mr. Glasso. Let me see how this wound was taken care of. Hanna walked over to Mr. Glasso. As he unwrapped her arm, she groaned in pain and nearly fainted. I am sorry, Hanna. I need to see this. The wound had been taken care of and was bandaged up, but Mr. Glasso knew that it could be to cover up what really happened.

What happened, Hanna? Did you see this, Mr. Parris? No sir, I was not in the kitchen. I was still in my office. What happened Hanna? Please, Mr. Glasso not in front of them Hanna whispered, keeping her head down. Mr. Glasso knew that Hanna was having a hard time. Let's go into your room Hanna, and you can talk to me there. They both went into Hanna's bedroom. Now tell me what happened. I only know she threw grease on you.

Will I get to leave here once I tell you? She thought. Please, I don't want to stay here any longer. Of course, Hanna, we have made arrangements for you to leave today with me and the police who are waiting outside. I need to find out what happened before we can press charges against the Parris's. She has been mean since I got here. She acts like she is jealous of me, and Mr. Parris is creepy, and I feel him watching me at night.

He acted like he was going to touch me in his office, and when he blamed it on me, she was upset and took it out on me by throwing the grease on my arm! What do you mean Mr. Parris was going to touch you? He asked me to go into his office and sit down next to him, and watch the game. It was just the way he said it. Did he touch you, Hanna? No sir, but if Mrs. Parris had not walked in, he would have. He bought me this tennis dress that is really short, and he keeps wanting me to put it on for him. I think Mrs. Parris thinks something is going on with us. Please don't leave me here.

"Okay, this is enough. Get your things," said Mr. Glasso. Hanna got her pajamas and Mr. Glasso started going through her drawers to get the rest of her things. "Where are all your clothes?" "This is all I have sir." "They never bought you clothes?" "I have one other outfit, and it is in the wash sir." "What are these?" Mr. Glasso took the bag of paper dolls out of a drawer and held them up. "They are my dolls sir. I make them. Please don't throw them away." "I think I saw these in the back of my car when we arrived here." Mr. Glasso gave them to Hanna without question, as he wondered why they meant so much to her. They were the group of brown paper sacks that he thought were garbage when transporting Hanna to the Parris house, but with more detail to look like clothes. He could see she made the sacks into paper dolls to coincide with the dolls.

Hanna got her toothbrush, toothpaste, and her one pair of pajamas as well as her paper dolls and held all of it in her arms. Mr. Glasso, are we leaving now? Hanna, we need to put your things in something. Where are the dirty clothes Hanna? We need to get your other outfit. I don't know sir. I just laid them on my bed when I took them off, and they were gone when I got out of the shower. They did buy me pajamas and decorate my room in Sunflowers for me. That was the only nice thing they did for me. They just do not like me. Let's get you a plastic bag to put your things in. Mr. Glasso walked out of the room and asked Mrs. Parris for a plastic bag, gave it to Hanna, and sternly motioned her to follow him out of the room. "Come on, Hanna let's get you to the hospital and have that burn looked at.

They walked back into the kitchen, where Mr. and Mrs. Parris sat. I am taking Hanna to the hospital and she will not be back. Mrs. Parris started yelling, "You are blaming us, and she is a little whore, and God only knows what would have happened with my husband if she were allowed to use her deceitful tricks!" You need to be prepared for one of our investigators to talk to you. Investigators! Well, that is fine, we have nothing to hide. Mr. Glasso put his hand on Hannah's back and led her in front of him and out the back door. Outside there stood a police officer, but there did not seem to be a need for him as the Parris's seem to be non-conflicting as of this time.

Mr. and Mrs. Parris now stood in the kitchen watching Hanna and Mr. Glasso walk out the door, get in their car, and drive away with the police following. Mrs. Parris was glad to see Hanna go. She did not care for her from the moment she met her and had it in her head that she was trying to move on with her husband. Everything bothered her, and Hanna could do no right. The Parris marriage was not the best, and Mr. Parris was not giving her the attention she needed. She had been his everything and he was now always looking at younger women and did not even try to make love with her anymore. They stayed married because they had been together so long, and there was so much equity in the farm, and the value was skyrocketing year after year.

Mr. Parris knew Hanna felt uneasy around him, and that turned him on even more. He had never had relations with a child, but the more he was around Hanna, the more he wanted her. He was able to see her confidence and positive attitude in herself, and he really liked that about her. Mrs. Parris was always so negative, and he was not attracted to her appearance anymore. He found it fun to play games with Hanna and see her squirm.

He could not wait to get her alone one day and get in her pants, and he had his plans in the palm of his hand when Mrs. Parris ruined them. He knew that one day it might come down to threatening Hanna, and the thought excited him. He wanted the excitement that he was lacking in his marriage, and he was going to get it one way or another. His lucky day was not about to come because the plans he and his wife both planned had fallen through, and now they waited for the investigator, uncertain if this would end with jail time.

CHAPTER: 14

Death Under the Bridge

Trisha had finally reached the bridge and was looking for the fix she needed so desperately. "It's cold out here. Do you mind if I join you all?" A group of homeless people were surrounding a burning trash can to keep warm, and couldn't care less that Trisha was there. They were all high and talking nonsense. Trisha was cold, exhausted, and downright mad. "Does anyone here even speak English?" Trisha said sarcastically. "What about a cigarette? Hell, I'll smoke the butts if someone just gives me a light." The man standing next to her cleared his throat and spit on the ground and grunted, as he stared right through her. Trisha made a disgusted face, rolled her eyes, and looked away. "Dammit! I need something to help me get through this night, and you guys know what I mean," she yelled.

Trisha stood in front of the fire, rubbing both her hands together. She was shivering. "I have three little girls," she said to the people standing next to her. "They are beautiful, and they are waiting for me to come get them. It will be soon, you know, as soon as I get a job and a place to live. I have plans, you see, big plans, and we are going to live in a beautiful white house with a picket fence one day. Four bedrooms, and a big kitchen with a window over the sink! My house will have three bathrooms and a huge backyard with a swing set and trampoline, and a couple of dogs. My little girls know this, they do. They are going to go to the best schools and have the prettiest clothes, you know. Everything they want I will give them." Tears were rolling down Trisha's cheeks as she noticed everyone ignoring her and staring at the commotion across the street.

Trisha looked across the street and saw a black car, an older Cadillac, drive up and turn off its lights. A few women who were standing around smoking cigarettes and drinking beer walked over to the car and got in. The car started up and drove off. Trisha knew these women, and they were hookers. She decided to walk in the other direction and pretend she knew nothing of them. "I may sell myself for drugs, and I was wrong to use my daughter as bait, but I'll never work for a pimp!"

As Trisha was walking into the shadows behind the bridge, a guy with a red robe, gold sash, and long, brown greasy hair tapped her on the back. "Hey, I heard you over by the fire, and I know what you are looking for. I think I can help you." "I don't have any money, so don't ask me for any." Trisha moved his hand off her shoulder. "Hey babe, if you come with me around the corner, I will take whatever you've got." "Show me first," said Trisha.

The guy in the red robe pulled out a bag of white powder. "This man is what you are looking for, hun." Trisha knew it was the meth she needed so badly. "Okay, where do you want to go?" "Just around the corner." Red Robe motioned Trisha to follow. Trisha followed the Red Robe as he went around the corner. Then Red Robe got into his pocket and pulled out the packet. He held it up, swinging it back and forth. "Take your clothes off." "No way!" said Trisha. "Give me some of that first!" Red Robe pushed Trisha to the ground, kicked her in the side, bent down and punched her in the face.

Trisha staggered to her feet and shoved Red Robe. He stumbled backward, screamed, and hit something sharp. Trisha looked around and saw that his back was bleeding badly; he must have hit something sharp as he fell. Trisha bent down and got into his robe, and took the bag of meth. "I am sure you deserved this." Trisha disappeared into the shadows. She found a large rock and snorted the entire bag of meth. Trisha was now high and definitely had her fix.

She felt really good and so good that she forgot the red robe was lying in the dirt, bleeding to death. Trisha noticed there were now two cars across the street. MMM, she thought, the Cadillac was gone with the girls, and now it is back. "I want a ride in the car!" Trisha stumbled and started walking towards the car, and another maroon car spun around the corner and started shooting at the Cadillac. Trisha had nothing to run behind and got caught in the crossfire and was shot twice in the stomach and once in the heart, and was killed instantly.

The people around the trashcan fire barely flinched. To them, it was just another night under the bridge.

CHAPTER: 15

News of a Mother's Death

Sami had been released from the hospital and was given doctors' orders of more rest and adding some more real food to her diet. She was to have the feeding tube changed regularly and was on her way to recovery.

The phone rang and Emma answered it out of breath. She had been outside throwing a football back and forth to the girls, and keeping Sami from running was like holding down an ox! "Hello? May I help you?" Emma answered, still out of breath.

"Hello, may I speak to Emma Black?"

"This is Emma Black. How can I help you? Who is this, please?"

"This is Sergeant Gates with the Midville Police Department. Do you still have Sami and Lindsey Flemming, foster children, in your home? You are their foster mother, correct?"

"Yes, why are you calling me, please? What can I do for you, sir?"

"Well," said Sergeant Gates, "I unfortunately have some bad news. There has been an accident. Trisha Flemming, the girl's biological mom, has been involved in an accident, and she is dead."

"Oh my God," Emma whispered, her breath catching. "How did this happen?"

"There are a lot of details that we are trying to sort through and get ironed out, but I know the body was identified and notifications have gone out to any family members that we have a record of. I am aware of these girls and their mothers' addiction problems. I am the original placement for Sami and Lindsey. They do have an older sister, you know?" said Gates.

"Yes, sir, I am aware of this. I am not sure what to do. These girls are so young," said Emma.

"You don't have to do a thing; social services will handle all details and will be in touch with you."

"They won't take the girls from me, will they? They love it here!"

"Well mam I can't say for sure, but DCS tries not to move children who have adjusted to their placement unless they absolutely have to, but be aware it can happen."

"If you want my opinion, Sergeant Gates?" said Emma.

"Yes, ma'am, Gates is the name."

"If you want my opinion, something was bound to happen to Trisha Flemming. She was a drug addict and a downright whore."

"Well, you don't know what her life was like growing up. It is not fair to blame addicts or their actions when we don't know their past," said Sergeant Gates.

Emma had no time to listen to any more, because she knew the girls could be taken away at any time. She was so fond of them and planned on adopting both of them. "Thank you, Sergeant. I will figure out how to tell the girls." Emma hung up and sat frozen, staring into space. She thought about how hard it would be to tell them and decided to wait until the next day, when it was morning and the day had not brought so many trials and tribulations.

Sergeant Gates knew that the girls would most likely stay with Emma Black and her husband. His worry was Hanna. He knew she was already moved to another foster home, and there was an investigation going on with the Parris's. He knew Hanna's history and how her mother pimped her out and how she was expected to take care of her sisters. How much more can a child take? he thought. "The three girls need to be together."

CHAPTER: 16

Fractured Justice

Sal attended Elkwood Junior High, a public school in Clark County. Most of the children were from large cattle ranches and well-to-do families. There was no private school in Clark County, so this caused a mix of children from all walks of life. If Hanna had stayed with the Parris's she would have attended Elkwood.

School had always been easy for Sal, and he participated in every sport he could get his hands on. Everything always came easily for Sal. Everyone loved him, as he was a very popular boy in school, and his future was promising not only because he came from a prominent family, but also because he worked hard and cared about achieving good grades and making something of himself.

Sal couldn't stop thinking about Hanna. His dad had said Lindman told him everything, and it was fine and Hanna seemed ok. Sal's dad had gone to the barn to check on Hanna, and she never uttered a word, seeming calm as a cucumber. They never heard anything more from the Parris's or from Lindman. Lindman said he never saw her again after that day and figured she had been transferred to another home. Sal wondered where Hanna went and what happened to her.

The Department of Child Safety and the police were looking into the Parris case. Sal's parents were only notified because Hanna included Riley and Sal in her narrative as to what happened in the barn. The only explanation DCS offered was that Hanna was being moved to another home and that Riley was under investigation.

"Girls, please take a break from the cookies and go watch TV for just a little while! I need to talk to Hanna about something." Hanna's foster sisters rushed into the living room and turned on the TV as Mrs. Nickel told Hanna to stay put. "Hanna, sit down," said Mrs. Nickel. "I want you to listen to me, okay? Sometimes things happen that are out of our control… sometimes bad things, and we have to be strong and get through them." Hanna looked at Mrs. Nickel with confusion. "Hanna, honey, it's your mama, something terrible has happened, and I am trying to figure out how to explain it to you."

Hanna looked scared and unsure. "What?" said Hanna. "Um, your mama was hit by a car and, well, she did not make it, she is, uh, well, Hanna, she is gone." Hanna looked at Mrs. Nickel and stared at her face for a very long time. "Dead?" Mrs. Nickel stayed silent but nodded her head yes. "What will happen to me and my sisters now?"

"I don't know, honey, but I promise you will stay safe."

"We don't have a grandma or grandpa, we don't know any family or aunt and uncle. My mom kept us away from them, I think. What will happen to us? I don't even know where my sisters live." Hanna began to cry.

"They took them from me and I don't know, I just don't know," cried Hanna.

"We are going to call your case worker and find out, okay? I promise, Hanna," said Mrs. Nickels.

"My sisters and I need to be together, please help us," said Hanna.

Mrs. Nickels felt horrible and did not know how to respond. "Do my sisters know about our mom?" Hanna asked.

"I am not sure. I am sure they will soon if they don't know already. I am waiting for the Sergeant to let me know some more information."

Mrs. Nickels did not know how anything would even be handled if there was nobody to handle it. "Would there even be a funeral?" she thought.

"Mrs. Nickels, you know what happened to me? This bad man needs to be in jail for hurting me. He does not deserve a good life because he may hurt other little girls. My mom was not bad; she took drugs and drank beer and did some bad things, but she tried to take care of us the best she could."

"Probably the best she even knew how," said Mrs. Nickels.

"I tried to help her and do what she wanted me to do, but I got so tired, Mrs. Nickels." Hanna began to sob as she hung her head down.

"I am not sure what all happened to you, Hanna, but I know it was not good. I am here to listen if you want to tell me." Tears slowly rolled down Hanna's cheeks. Hanna nodded and said, "Mrs. Nichols, can I still bake some cookies? I don't want to talk about this right now."

"Sure, honey, call the girls back to the table." Mrs. Nickels slid the bowl of cookie dough over to Hanna.

Hanna and Mrs. Nichols' girls made cookies that day; however, Hanna was very upset inside, and once the girls got tired of the cookies, Hanna quietly went into her room and got out some of her paper dolls for comfort.

The Parris's barn was now surrounded by police. Sergeant Gates knocked on the barn door. "Lindman?" There was nothing but silence. "Lindman!" Gates raised his voice and yelled louder. "This is Sergeant Gates from the Midville Police Department. I need to talk to you, sir."

Lindman hesitantly walked out of the back of the barn. "I am sure you do," said Lindman. "Hanna, right? Sal just has to cause trouble again, that boy. He is upset because I won't pay him any more than he is worth, and let me tell you all how I believe, you gotta do a great job to get great wages, and that ol Sal, well he likes to mess around and daydream about girls and basketball, you all know. Parris's only give me so much money to distribute to the hired help."

"I need to ask you a couple of questions. Where were you on the afternoon of June 12th?"

"I was here, and you know it, Sergeant. All I do is work!"

"Who was with you?"

"Sal, and Hanna, that foster girl up at the Parris's, she is a real piece of work, believe me, and ol Riley stuck his nose into everything that afternoon too! You know damn well I was here, Sergeant, and with Riley and his boy Sal being here as well, there is something bound to happen with those two. So you tell me what you are doing here before I just decide to end this conversation real quick."

"Lindman, some serious accusations are being said about you."

"Now you are pissing me off. I told you that Sal is upset because I don't pay him enough for his little boy stunts he pulls!"

Sergeant Gates' face started turning red, and he felt enough anger in him to explode. "No! It is Hanna, she is saying you raped her!"

"For God's sake, Sergeant, you know me, I would never do such a thing."

"Why would Hanna make this up? There is no reason for that girl to lie, and you like the women, Lindman, everyone in Midville County knows it!"

"Other men like women too, Sergeant, it is a common thing among us men, you all know!"

Lindman laughed and shook his head in disagreement. "You don't have proof of anything, Sergeant, only what that girl is telling you all, and she is some foster child that is seeking attention, and you know it!"

"We are going to be looking into this further, Lindman. We have a whole lot of questions for you, starting with the ones that need to be answered down at the station."

"Now?" asked Lindman. "I am not finished with my work here at the barn, and the Parris's will hold my pay, and there are others involved in running this ranch and depend on the money I give them!"

"Let's go," said Sergeant Gates. Gates was fed up and knew Lindman had done something he should not have.

His personality got him in trouble daily, and there were past issues that were never completely proven, and now was the time that Gates was going to make sure Lindman was arrested for all of it. He knew Lindman raped Hanna without even knowing the entire story.

"I will drive myself. I am not under arrest yet."

Gates nodded. "We will be happy to follow you, Lindman," and he motioned his men to get back into their vehicles. The men all backed away from the barn and got into their police vehicles as Lindman pulled out his truck and headed towards the station.

CHAPTER: 17

The Day of Telling

Emma Black stood at the kitchen sink, eating a donut and staring out the window, trying to figure out how to talk to Sami and Lindsey about their mom. Abi, their caseworker, had called that morning and told her what had happened to the girls' mother. Abi said she would talk to them, and Emma pleaded for her to let her tell them. Abi agreed but asked her to keep it brief. They were still little girls. What would she say? The girls called her mom now, and she knew foster children did this as she learned it in her foster care licensing class, but her love for them had grown deeper every day, and this was going to be hard.

Another donut sounded good, so she picked a chocolate one from the box on the table. "Mom! Sami is making me crazy!" Lindsey exaggerated. "She put her playdough in my teapot, and it got hard. Now, it won't come out, and if I try to make it come out, it is going to break the teapot, Mommy!" Sami ran past both of them through the kitchen and stuck her tongue out at Lindsey. As she ran through the kitchen, she tripped over a rug and went flying into the cabinet. The cabinet door was open, and as she went to catch herself, her hand caught some glass bowls, and they flew out of the cabinet, shattering on the floor. Sami screamed in fear and started crying hysterically, as Emma ran over to Sami, picking her up and hugging her. "Girl, this is bound to happen when you are running through the house. I have told you girls time and time again not to run in the house."

Sami darted to the coat closet, opened it, and shut herself inside. "Lindsey, what is she doing?" "I think she is hiding, Mama. Why would she hide from us? It's okay, Sami, come out of there." Sami began to cry and scream louder. "Stop trying to get her to come out, mama, because she is scared. My mom's boyfriend used to hit her when she did wrong, and Hanna and I would hide her in the closet so he would stop." "Oh my gosh, Lindsey. I had no idea." "Sami, honey, please come out of there," Emma knocked softly on the closet door. Sami continued screaming. "Let me help," said Lindsey. Lindsey began singing a song that her mother would sing sometimes. It was a German song that their grandmother taught their mom when they would go fishing. Lindsey began to sing quietly, and Sami began to quiet down and stopped screaming.

"Mama, you can talk to her now." "Sami, honey, nobody is going to hurt you. We love you, please come out so I can hold you, honey and make sure you are ok." Sami slowly and hesitantly opened the closet door and crawled out to Lindsey. Lindsey bent down and put her arms around her. "It is ok, sissy. Mama is worried about you, hug her." Lindsey called Sami sissy at times when she needed affection. Sami crawled over to Emma and crawled into her lap, hugging her. "Sami, I love you," said Emma, "You don't ever have to be afraid. Things break by accident, and you would not have been in trouble for breaking those dishes. Even if I did get upset about my dishes being broken, I would not have hurt you because I was mad. You don't have to hide in the closet. You should have no fear here in this house, honey." Sami just kept hugging Emma, but she was completely calm now and listening to what was said.

"That's it, I have to tell them now," thought Emma. "Come on, girls, let's go into the kitchen and sit down and talk for a bit. This has been a lot to deal with, and I think I will continue on with some more conversation you need to hear." Emma really did not want to tell the girls as of this time, but she did not want to go through another day of trauma and screaming, and decided to get it all over at once.

The girls followed Emma into the kitchen, asking no questions as the episode with Sami was now in the past and things were finally peaceful. Emma slowly sat down on a kitchen chair and quietly motioned the girls to come over to her. Sami took a deep breath as she was worn out, and walked over to Emma, and Lindsey followed. "What's wrong, Mama?" asked Lindsey. "Sami is ok now; she isn't upset anymore. She'll get better in time, I promise." "No, honey, it is not Sami." Both girls were now standing in front of her with anticipation, as she started to explain what she thought would be the hardest conversation in her life.

"Girls, I know we just went through a little bit of chaos with Sami and the closet ordeal; however, I need to talk to you about something very important. Um, girls, I know you are young, very young, and I know life has thrown you lemons and just downright crap, a bunch of crap. I want you to know that we love you so much. I also have never seen two sisters who love each other so much." "You mean three, Mama?" Lindsey said. Emma started to cry. "When you were placed in my home, I just loved the fact that you had each other, and I loved that you were so close in age. I knew you had an older sister, and in my heart, I wanted to find her somehow and bring her here. I know time has moved on now for a while, and we still don't get to see your sister. I just want you to know that I do inquire about her, and DCS is aware that we would love to have all three of you." "Really?" asked Lindsey. "You mean she can come live here too?" "Yes, if DCS will allow her to." Sami jumped up and down, clapping her hands. "Wait, Sami, we can't trust what will happen yet," said Emma. "Let me find out more. I don't want to get our hopes up only to have everything fall through."

"Girls, before me, you had a mommy. Do you remember her?" Emma looked at Lindsey for an answer. "Yes, she would fix us hot dogs." "And she loved doggies," said Sami. "I remember she would cry a lot at night and argue with Hanna. Hanna would cry too, back in mommy's bedroom. Hanna did not like mommy's friends," said Sami.

"Girls, something has happened to your mommy. She is up in heaven with God now." (Emma hated feeling like she was lying, but she was trying to make things easier.) "Um, there was an accident, and God called her home. She is gone, girls, she isn't on this earth anymore and won't be again. You won't see her again." Emma wiped her tears away. Sami and Lindsey stared at her, expressionless. "I am so hungry Mama, can we have macaroni and cheese?" Emma sat in silence. She had never expected a response that was so desolate and lifeless; it was killing her heart.

"Yes, girls, macaroni and cheese it is." Both girls ran off into the family room and started jumping on their indoor trampoline. Emma bowed her head. "Dear Lord, give me the strength to handle today. We both know my husband and I are too old to raise these children at their young ages. I don't know what to do. God, they need us so much. I can't allow them to leave here and go on their own into more darkness, another foster family, or adoption of people who don't understand and don't genuinely love them. God, they don't even feel the pain about their mom. How could they not feel God? Emma dropped to her knees. Father, strengthen my heart to fight for what is right in Your will. In Jesus' name, I pray."

"Mommy, why are you on the floor?" said Sami. "I was just saying a little prayer." "Mommy, we are hungry!" "I know I am fixing it right now. Let's have some strawberries with that mac and cheese, OK?" "Ok!" the girls yelled. The girls ran back to their trampoline. "I am fixing it right now!" Emma chuckled, and as she got in the cabinets to pull out the box of macaroni and cheese, she looked up and smiled and said, "Thank you for giving me the courage, Lord, because I have none. I know you gave these children to me for a reason. I have to continue to take care of them and love them. I know my mission Lord, it is my calling, thank you, dear Jesus."

Emma finished the macaroni and cheese and cut up the strawberries. "Girls! Come and eat! … Go wash up." Sami stomped her feet and ran to the bathroom. "I'm first!" she yelled. Lindsey ran behind her, "Whatever, I am faster anyway." Emma felt a bit of peace for now, but was also worried about how she would tell them about their mom when there was no emotion. She decided not only to turn it over to God but also to call their caseworker, Abi.

The girls finished their lunch, and Emma told them to go play outside for a while. After cleaning up the lunch dishes, she opened the back door and told them to come in. "Girls, come in and let's get out the paint and paint awhile." The girls came running in. "Where is the paint mama?" said Lindsey. "It is in the hall cabinet." Lindsey ran to get the paint and came back with a box of paints in many colors, brushes included.

Emma put a plastic tablecloth down, and the three of them painted for a few hours.

"A good mother is a gift to a daughter." Emma thought. She was happy that she allowed her faith to step in on this day and calm her down. She was determined to give them a life they never would have had and spoil them with all the love that she and her husband had for them. Age was not a factor anymore; she decided they were destined to be a family. Staying healthy was the name of the game now, and she regretted eating those two donuts earlier. "Starting tomorrow, I'll eat healthier and begin an exercise program so I can be here for these girls."

The three of them finished painting and put their paintings on the washer and dryer tops to dry. Lindsey painted a sunset with smiling faces, and Sami painted scribbles and was learning to make a star, so they were all over the paper in many different colors. Emma decided to call her husband to pick up a baked chicken, "that's a healthy start to my plan", she thought, a baked chicken.

CHAPTER: 18

Under Interrogation

Lindman pulled into the police station along with the other patrol cars. Sergeant Gates parked his vehicle and motioned the other officers to do the same. Lindman walked up to Gates and said, "What now?" "Have a seat inside." Lindman opened the front door and sat down in the waiting room as Gates followed. "Not there, buddy, follow me," said Sergeant Gates. Lindman got up and followed Sergeant Gates into a side room. "Have a seat, Lindman." Sergeant Gates pulled out a chair for Lindman, and Lindman sat down. "You can't prove anything," said Lindman. "When a child describes in detail when, what, where, and why, I honestly believe there is some truth there, cowboy."

"I don't have to put up with this shit!" said Lindman. "Someone is trying to frame me. We all know Glass pays really well, and my job was wanted by many." "Lindman, I want to go home tonight," said a tired and disgusted Gates. "Are you going to give me a confession or not?" I want an attorney. No problem, Gates said. "We have Hanna's statement." I don't want to have the girl testify in court, but I will if that is what it takes to get thugs like you off the street. Now, Gates, I thought you liked me, man? Lindman said with a sarcastic look. Lindman, you are a real fool.

Gates motioned one of his deputies to come into the room. You have the right to remain silent, you have the right to.

Lindman started yelling, "What the hell are you doing, Gates!? Sergeant Gates handcuffed Lindman as he resisted. "You have no proof!" said Lindman. You are taking the word of some poor foster kid who goes from home to home, has a crappy life with parents on drugs! Sergeant Gates was done and completely stressed out. We have enough information on you, Lindman. I just read the medical reports. Hanna had a forensic exam, a physical exam, and a pregnancy test. I will say the Midville Police Department and the hospital work very well together. And for you, my man, we're scheduling a swab test and collecting DNA samples, as well as conducting forensic examinations for further evidence. Lindman started laughing and got louder and louder, until he almost sounded insane. I have wondered about you for years, said Gates, and I could not put my finger on it, because I will admit you are pretty damn slick. You still don't have proof, Lindman spat at Gates. I have had worse than you, don't forget it, Lindman, you are free to go. Sergeant Gates uncuffed Lindman. When we get the results, you will be back here where you belong. Sergeant Gates left the room.

Sal's Dad arrived home after checking out the situation at the Glass's house. Sal met him at the door. "Dad?" "Well, something is going on for sure. I tried not to let anyone see me because it really is not my business." "Dad! What? What?" said Sal nervously. "Look, Sal, I saw Lindman talking to the deputies and then I saw him get into his vehicle, and all the deputies got in theirs. "Oh crap!", said Sal. "Now Sal, just wait a minute and let me finish, quit being so impatient and judgmental!" Sal looked disgusted and rolled his eyes as his Dad tried to finish what he wanted to say. "I followed them and they all went into the police station. I waited an hour and Lindman walked out, got in his car and left. "Are you kidding me, Dad? They did nothing?" "Now Sal, we have no proof of anything! I know Lindman is in trouble; there is a lot at stake and a lot that needs to be proven before he is arrested. "Why don't they just ask Hanna," said Sal. "She was kind of there!" "I am sure they are, son, I am sure they are."

CHAPTER: 19

The Funeral

"Sit still, girls. I know it's hard for you to understand this, but you need to be quiet and sit still." Emma felt horrible about the funeral, but relieved it would be over that morning.

"We are," Lindsey said, a little sassy. "We know what's going on. We're not dumb, Mama."

"We are," Sami echoed, copying Lindsey.

Emma knew Lindsey understood what was happening because she was a bit older. She could see that Lindsey didn't want to accept the hurt or acknowledge her anger; instead, she was pushing it all to the back of her mind. She also knew Sami was just mimicking her big sister because she was little and could. No matter what, it was a hard day for all three girls.

The funeral was held outdoors as a simple graveside service. There weren't many people, and everyone was dressed in black. Most of those standing around the casket were from county agencies, the sheriff's department, foster care workers, and attorneys.

"Hanna!"

Emma's head snapped up. She immediately recognized Sami's voice calling out to her sister. Emma shut her eyes for a moment and whispered a quick, silent prayer. *Father, please surround my girls with Your love and keep them from getting hurt and confused today. Thank You, Father.*

Sami threw her arms around Hanna and began to cry, her tears quickly turning into deep sobs. Lindsey saw Sami running and followed, spotting Hanna in the distance.

"Hanna, Hanna, Hanna! Oh my God, Hanna!"

All three girls, Hanna, Lindsey, and Sami, wrapped their arms around each other, weeping with joy as if they had never been apart. They started laughing through their tears and jumping up and down, clinging to one another.

Emma watched this outpouring of love and thought of two of her favorite quotes, one by Herbert Hoover, reminding her not to rush the girls today and to let them have their time together, and another by Fyodor Dostoyevsky, reminding her that, no matter what happens, children still see a magical world in front of them.

– Herbert Hoover –

"The extra hours you put in today will keep a smile on the faces of your children a few years from now."

– Fyodor Dostoyevsky –

"Children see magic because they look for it."

Hanna had come to the service with her current foster mom, Tami Nickels. Tami smiled as she saw the girls hug each other and get so excited. Tami walked up to Emma and was going to start a conversation, but Emma started one first. "The girls are happy with us."

"And you are?" said Tami.

"Excuse me, I don't think we have been introduced. I don't know you, do I?" Emma felt sick to her stomach. There was so much happening in the last few weeks, she felt like she could not keep up.

"I'm Hanna's sisters' foster mom."

"Oh gosh, I am so pleased to meet you. Hanna does nothing but talk about her little sisters. This is such a sad time for her to see them," said Tami.

Emma nodded in agreement, "At least they can see each other until one of these agencies pulls them away. Good old Abi is eyeing us right now from DCS. I know they mean well, but it just gets old with the in and out of everyone in your home. I hate it," said Emma.

"I totally agree," said Tami.

Hanna looked up at Tami and began to speak, "These are my sisters, Lindsey and Sami," Hanna said with a very loving smile. Hanna looked as though she were their mother, with the way her love shone through her eyes.

"Hello," said Tami with a grin.

"And who is this beautiful, tall young lady?" said Emma.

Sami giggled as she said, "This is my sister, silly!"

"It's Hanna, my big sister," said Lindsey.

"Oh my goodness, it is so nice to meet you, Hanna," said Emma.

Sami and Lindsey giggled and jumped up and down some more, but Hanna remained silent. She was older, and she knew that this happiness was only temporary.

Hanna slowly slipped away from Lindsey and Sami and walked up to Emma and Tami. They both smiled as she walked up. "Why can't my sisters come with me?" A tear rolled down her face as anyone could see she was trying to be a brave little girl.

"Unfortunately, honey, things just don't work this way. We can't just make the decisions of what will happen here today. As much as you are loved by us and as much as the Blacks love your sisters, we can't make the decisions."

"Who makes the decisions?" asked Tami.

"What do you think, Emma, about maybe having a play day with the girls soon, today?" said Tami.

"Hanna, why don't you run and play while Tami and I discuss something quickly," said Emma.

Hanna was upset, but she listened and ran back over to her sisters with a half-hearted smile on her face. She grabbed both of them and hugged them and pulled them into her lap as she sat on the ground holding them, how she always had when they lived together at home. The girls sat in her lap, looking at her attentively, listening to her every word.

"Is there a problem with the girls getting together, Mrs. Black? I can set it up. DCS supports sibling bonding when possible."

"How is Hanna doing?" said Emma.

"Hanna is great. After all that girl has been through with her mother's death and her prior foster family, that girl is one tough child. She gets along with my girls wonderfully," Emma said. "Has she been to counseling?"

"Yes, she has, and she seems to be doing wonderfully," said Tami.

"This is what I question: the trauma she has had. I don't think it is good for her sisters to be a part of that, when they are doing so well with my husband and I now," said Emma. "It could really mess Lindsey and Sami up if Hanna talks about any of her prior foster experiences."

"Yes, we both know what that entailed," said Tami sarcastically.

Emma looked puzzled. "How do you even know about Hanna's case, Tami?" said Emma. "DCS does not share the case information with anyone else but those who are involved in the case itself."

"You know how foster parents talk in this community, everyone knows about everything, and we all know what happened to Hanna and who did it! I am sure you know everything because you have her sisters. Actually, Tami, you can stop right there. Hanna's prior foster family and their involvement in anything should be confidential, so please do not pre-judge and come up with your own conclusions that can be very harmful in the long run."

"I am not trying to cause a problem here, Emma, I just want to get the girls together to play sometime," said Tami.

"I can't have my girls scared by what has happened to Hanna. I am not even sure of all the details; it is too risky." Emma was done with this conversation.

Tami was not going to give up. "So you want to keep Lindsey and Sami away from Hanna when it was not even her fault? Hanna did not ask for horrible things to happen to her; she did not ask to be taken into the system, raped, and abused! She has been through the worst of the worst, and if her little sisters could understand, which they will one day, they would want her to be with them!"

"You have to be kidding me, right? said Emma. You seriously have a problem understanding humanity! I understand you want Hanna to be a part of my girls' lives, but as long as they're with my husband and me in foster care, we're responsible for them, and I don't think staying any longer than we have to is right right now."

"What? Right now? Her life is not going to go away, Emma! She needs her family, and she loves her sisters. You take them away for good, and she will be devastated. God only knows if she will ever even recover! cried Tami."

"I am sorry," said Emma, "I did not mean to upset you. I need to go." Emma started walking away.

"You are so very wrong, you are so very wrong, and I will talk to DCS about this. These girls need each other, and you know it!" yelled Tami.

The service was ready to begin, and everyone was gathering around the casket. Emma was not sure who was in charge of the service, but she was glad there was one, outside of wanting to take the girls away from it all before it even started.

Emma walked up to the girls. "Okay, Sami and Lindsey, we are going to go soon, so hug Hanna."

"Nooooo," the girls whined. "You are leaving now? Why? You are not staying to see the service? Hanna, honey, you need to go find your foster mom, ok sweetie? Are you leaving? When will I see my sisters again?"

Emma took the girls' hands and led them to Hanna. "Hug Hanna and kiss her on the cheek."

Hanna bent down so they could reach her cheek and gave them both a big hug. "You are leaving now, aren't you? It is for the best, Hanna; your sisters don't understand they are both so young said Emma."

Hanna did not say a word and stood silently as Emma took the girl's hands and walked away. Sami and Lindsey turned around and waved goodbye to Hanna. Hanna waved back and yelled, "I will see you soon!"

Hanna hung her head down and walked back over to Tami. All she could think about was seeing her sisters again, and finding out why Emma left so fast. Tami hugged Hanna. The service had started, and Tami and Hanna sat down to listen, but Hanna was not focused on her mother's death whatsoever. She had become emotional when she was originally told about her mother's death, but her sisters were her concern, not her mother. She had never asked what exactly happened and how she died. Abi, her caseworker, had talked to her, but Hanna did not show any emotion at that time either. She just asked how soon the funeral was and if she would see her sisters. Hanna had taken care of them since she could remember, and most of the time her mother was gone or in the bedroom with her guests.

The graveside service began; the pastor spoke, and "Wind Beneath My Wings" by Bette Midler played softly across the cemetery. There were only about twenty-five people at the service, and Hanna only recognized two of them. One was her caseworker, Abi, and the other one was her mom's last boyfriend, Rome, who was arrested when they were taken into foster care. Why was he there? she thought to herself. Hanna did not know him well, and what she did know, she did not like. Her mother tried to get her to sleep with Rome to get drug money, and Hanna always found a way to get out of it, so at least she did not have to feel ashamed of herself seeing him.

A few people got up and spoke, and Rome was one of them. He said he was sorry for the loss and that he loved her very much. He said what a great woman Trisha was, and that aside from their problems, they were planning on getting married. Rome added that he would love to support any family members who needed his condolences, and that he created a GoFundMe account to help Trisha's children who were placed in foster care in the future.

"What is a GoFundMe account, Tami? This guy…"

Hanna interrupted, "You mean Rome?"

"Oh, you know him?" Tammy asked.

"He was one of my mom's boyfriends, and they were not getting married. She had only known him for a little while."

"Well, he is asking for money because he says he is going to give it to you girls."

"He looks like a real loser to me," said Tami. "We need to report him to your caseworker and let them call the authorities on him. Creep."

Tami was disgusted, but they listened to Rome go on and on, and then someone got up and read a poem. Hanna did not know this person, but it was a cousin of Trisha's that had never been mentioned.

The casket was lowered, and once the service was over, Tami and Hanna walked to the car, got in, and started to drive away, when they were halted by Rome as he stepped in front of the car.

"Hey Hanna, did you hear what I am doing for you?"

Hanna just stared at Rome and looked at Tami, and asked if they could go. Tami drove off.

"I am sorry, Hanna. I will call Abi tomorrow about Rome, but I am sure she understands completely because she was sitting right there listening to him. Sometimes there are just people who will do anything to make a buck, and they will use their loved ones, including children, to get what they want. He is probably desperate for money."

"Do you think he really might do it, Tami? Send me money? I could get my sisters and buy us a house, maybe."

"Honey, we have enough to think about besides buying you a house. We need to make sure you are always safe and figure out how to get you some visitations set up with your sister. I know you love them, and I promise you I will make sure I talk to DCS about all that took place today."

The day was almost over, and Hanna felt drained. All she could think about was playing with her paper dolls that would allow her to move into another world of happiness and forget every bad thing that had ever happened to her and her sisters. They were her imaginary world that she wanted so much.

Tami Nickels and her family began having trouble with her parents. They were elderly, and both were living in assisted living. Tami's mother's dementia took a turn for the worse, and she ended up being locked down in a dementia facility. Tami's father was distraught with his wife being so ill, and he started declining as well. Her father had made Tami power of attorney because they had helped take care of them for years, but when the Nickels decided to put Tami's mother in a dementia facility he had his other son Keaton take him to an attorney and he changed the will to exclude Tami and her husband, stating they were trying to take his house and steal all his money. He also stated that his wife had been verbally abused by Tami through the years, and that is why she got so sick. Even though these things were not true, they created a case with Adult Protective Services. DCS found out and decided it was too stressful for Hanna to live in the Nickels' home while their pending case was being figured out. They thought Hanna had been through enough, so she was removed from the Nickels' home and taken to another foster home against her will and that of the Nickels.

Hanna was going to be finishing her eighth-grade year at Junior High and was starting to have changes with her body, and she started her period. She was becoming absolutely beautiful with her long, auburn hair that fell into ringlets everywhere down her back. Hanna asked to see her sisters regularly and was always told DCS was working on getting that to happen. Emma fought against Hanna seeing Lindsey and Sami, and used the excuse that the girls would be scarred by her and who she was becoming, with all the trauma of the rape and her past. DCS was still looking into everything and seemed to be taking their time allowing visitation.

Hanna now had a sadness about her because she loved the Nickels and felt like she had no say in the matter of removal. She was getting tired of asking to see her sisters, but kept trying every chance she got. Even though she was getting older, she still found time to play with her paper dolls and made new ones when she could. She started to name them, and they became alive in her mind with names and personalities of people she wished she had met or people she would like to be like.

The home Hanna was taken to next was the home of Matthew Walters. He was a disabled vet with one little five-year-old daughter. Mrs. Walters had passed away a few years back and Matthew thought it would be a good idea to do foster care so that his daughter would have other children in the home to play with. DCS knew Hanna was older and going to be starting High School, and this may not be the best fit, but there were no other homes available at the time. They knew that no matter what home Hanna was in, she would try to make it work just so she could eventually live with her sisters.

The Walters lived in a low-middle-class neighborhood. Matthew was hoping for a younger child, but agreed to take Hanna because they were told by DCS that Hanna was a good girl and caused no issues. He also thought he could use some help with his daughter, Brandi.

"I really liked the Nickels, Abi. I felt really safe there, and I loved playing with the girls."

"It was not a good environment for you anymore, Hanna. I have told you this a few times now."

"It was too," said Hanna. "Nobody hurt me; they were busy with Grandpa and Grandma, but they still had time for us. Lincoln and Lanna were my best friends. I never had the best friends before."

"You will see them again, Hanna. This is a small town, and you will see them."

"Just like my sisters that you always tell me I will see?"

"Stop!" said Abi. "You don't know all that was going on there. It was too much stress for a young girl like you. You are a ward of the state, and the decisions are made by us, and this is for the best."

"You did not even ask me what I wanted!"

"Hanna, after what happened to you at the Parris's, we can't allow any additional stress or problems. I'm done talking about this, we're almost to the Walters' house."

CHAPTER: 20

New Homes, Old Haunts

Over three years had passed, and Sami and Lindsey were now adopted by the Blacks. Hanna crossed their mind now and then, but they had a new life, and it was very busy. They were both very happy and involved in everything, gymnastics, Girl Scouts, softball, theater/chorus, and Lindsey was learning the violin. The Blacks had an expensive lifestyle, but they could afford it. Canby was a Patent Attorney, and Emma worked from home as a Financial Consultant. They loved the girls so much, they wanted them to experience everything in life.

Sammi was now five Lindsey was six, and Hanna was thirteen, going on fourteen. Both Lindsey and Sami attended a private Christian school, and they were on the principal's list. They loved their school and they had many friends, which they had never had in their prior life with their mother.

"Your vests are all done, girls!" Emma had just finished sewing their Girl Scout badges onto their vests.

"Thanks, Mom," said Lindsey. "Cool," said Sami, grabbing her vest and putting it on.

"Ok, you guys have your Girl Scout meeting tonight at the church, and tomorrow you have a softball game, so let's find your uniforms and get them in the wash!" The girls ran to their rooms, pulled their uniforms from the hamper, and tossed them on the kitchen table. "I will never get caught up with the laundry with all of us running around all the time," she laughed.

Canby walked into the house and told the girls no one was going anywhere; he'd just been to the church and heard there had been recent thefts.

"Emma, people are taking things from the church parking lot during services. Honey, they can't go tonight. I have a bad feeling about what's happening over there."

"Nope. I'm sorry, girls, your Girl Scout leader doesn't lock the door once you're inside; I've seen it myself."

"For God's sake," said Emma, "I'll take them myself and stay with them tonight."

"Nope," said Canby. "I'll take them, Emma. Come on, girls, get your vests on and let's go."

"Wait!" yelled Emma. "Your teeth, your hair, you're a mess!"

The girls turned around and ran back to Emma. "Go!" Emma pointed to the bathroom as the girls raced down the hall to get there.

"You make so much out of things," said Emma. "They are my life just as yours," said Canby. "I need to talk to the Girl Scout leader and have her start locking the door."

"Please do, it is so dangerous," said Emma. The girls ran from the bathroom into the living room, kissed their mom, and out the door they went with their dad. Canby grabbed their hands as they crossed the street; the girls looked up at him and smiled. They were so happy, and they were turning out to be Daddy's girls.

Sal was outside his home, shooting baskets, when his mom came out with the newspaper. She was shaking her head, saying, "It's about damn time!"

"Sal, you will be happy to know Lindman was finally arrested, and so were Mr. and Mrs. Parris."

"What?" said Sal. "I was reading the news today, and it says right here, look right here." Sal's mom pointed to a section on her phone as she gave it to him.

Sal continued reading: "Yep, it says right here that Lindman got arrested for statutory rape, and look as you read down the page, Mom, it mentions Mrs. Parris for throwing hot grease on Hanna's arm and other abuse that they do not mention. They arrested Mr. Parris as well, probably because he was aware of everything and did nothing, don't you think, Mom?"

"Hell yeah!" Sal jumped like he'd just sunk a basket!

"I wish I knew where she was, Mom. I wish we could know she is ok."

"Well, she was taken by The Department of Child Safety, and I am sure they found her a good home."

"What? Are you kidding me, Mom? They found her a home, and look what happened to her mom?"

"Sal, they did not know that would happen! They have no way of knowing, son. They check out the foster parents, and sometimes things don't go as planned. They are now taking care of everything. You don't know that she is not okay, son. You don't know that at all."

"I wish you and Dad could become foster parents so that we can take her in, Mom."

"Look, son, it's been a long, long trial, and it finally has ended after three years. Hanna is probably adopted by now. Be happy for her that this turned out in her favor. We need to move on from this. We have other focuses right now, like you starting high school next year!"

"Sal, you are going to find out what it is to be successful, my son, and get scholarships offered to you in so many areas you can't keep up!" said Sydney.

"Mom, you have so much faith in me."

"Yes, because I know what you are capable of in Academics, as well as sports. The sky is the limit, Sal, it really is." Sal's mom smiled and grabbed her phone out of his hand. "We're done with this conversation, no more. If you want to help Hanna, wherever she may be, pray for her son."

"I will, Mom," said Sal, as he started shooting baskets again.

Hanna and her caseworker, Abi, knocked on the door of Matthew Walters. Hanna glanced into the front window and saw a little girl run through the front room. A man with long, brown, curly hair and a beard opened the front door, and the little girl was now seen hiding behind his legs.

"Brandi, get back, stop hanging on me!"

"Hi, I am Matthew Walters." The man held out his hand as if he were ready to shake Abi's hand.

"Hello," said Abi, not taking the offered hand. "I'm from the Department of Child Safety. You should have been called and told I was on my way. I received approval to bring Hanna here."

"Of course, come on in," said Matthew.

Abi opened the door for Hanna and motioned her to go in. Hanna looked up at Matthew and quickly looked down with embarrassment.

"'This is Hanna," said Abi. "Hanna is a little shy, but a good kid."

Brandi was still hiding behind her Dad. "She will warm up, she is not used to a lot of company and friends and all," said Matthew. "I am glad you are here, Hanna."

Hanna looked at Matthew with a sense of uncertainty and said, "Thank you."

"Here's her file, Matthew. If you have questions, just call the office; the number is on the first page."

Abi looked at Hanna and winked. "You will be fine here, oh, and here is your bag. It is right here by the front door." Abi handed the bag to Hanna and walked out the front door. Matthew locked the door behind her.

"So, Hanna, this is my little girl, Brandi."

Brandi had long, beautiful brown hair to her waist. Cute as can be. "Hi, ma'am," said Brandi. Matthew chuckled as he sat down on the couch next to Hanna.

"Um, you want me to show you your room? It ain't much, but it's clean. I cleaned it with Brandi's help. We just have to keep Brandi out of there; she's going to want to visit you."

"It's okay," said Hanna as she picked up her bag and followed Matthew down the hall into her now so-called bedroom.

"Here you go, hun. I am sure you want your privacy. See, over there is a bathroom," he pointed across the hall.

"We only got one bathroom, but I'll try to keep it clean for you all. Maybe you can help me."

Hanna smiled slightly and said, "Yes, I would like to put my things away if that is ok."

Brandi barged in. "I can do it, Daddy. Here, give me your bag." Brandi grabbed the bag and started to unzip it.

"There isn't much in there, little one," said Hanna.

"Brandi out now!" said Matthew. "Give her some space!"

Brandi looked up at her dad with a pouty frown and walked out of the room.

"We will talk more later, hun, I gotta get some dinner started for you girls," said Matthew.

"Thank you," Hanna said.

Matthew quietly shut the door, and Hanna started putting her clothes away in the drawers. They were old chests of drawers, and the drawers kept slipping off their tracks, but at least she had a place to put her things. She had a twin bed, and it looked clean, but the room was old, and the ceiling appeared to be in danger of collapsing at some point. There was no fence around the yard, and that bothered her because someone could come right up to the window and look through the thin material the curtains were made of. "I will get dressed in the bathroom," she thought.

While settling in, Hanna found the last batch of paper dolls she had created at the Nickels' and laid them on her bed. She sat down, took out some of the clothes she had designed from her plastic bag, and started dressing the dolls. "Here we are, girls, we are home once again, if you can call this home. I will not put up with anyone coming in here to hurt us. We will run from this house and keep running and never look back." I don't have to be happy here. This is only temporary. I will move my dresser in front of the door so nobody can hurt us. You can hide under the bed, or behind the shower curtain, and throw hot water on this foster parent! Don't eat the food he gives you! He could burn you with hot grease!"

Hanna was finally showing signs of emotional trauma and a decline in her mental health, all in her short life. Hanna began acting out with her paper dolls in such a way that it would seem scary to most. She stared out the window that day, really thinking of nothing, just staring. She slowly put the paper dolls back into the plastic bag and hid them in her bottom drawer, under her clothes.

The paper dolls Hanna had started creating when she was very young were helping her cope with her traumatic past. It was her own created therapy. Hanna had scheduled mental health therapy through the Department of Social Services; however, they were short on staff, and many sessions were canceled. When therapy did take place, her trauma would be discussed, but then after a few sessions, there would be a new therapist, and Hanna felt uncomfortable telling the deep, dark secrets.

It was summer, and school was around the corner. Hanna was going to be a Freshman, and her sisters should now be in kindergarten and first grade. Hanna missed her sisters but knew the system was in control. It was not that she was giving up on seeing them, but she finally accepted the fact that she was only 14 and that nothing she would say would mean anything to them.

Hanna adjusted to her new foster home at Matthew Walters' house, but she wasn't happy. She just accepted the daily routines. There was something that bothered her, but she couldn't quite put her finger on it. She saw a lot of anger in Brandi. There were times when, if she did not get her way, she would scream for hours. She also didn't understand why Brandi wasn't potty-trained at five years old. She wore a pull-up day and night, and there was no mention of the pull-up or trying to train her. Hanna's sisters were so easily trained, and their mom was always high or drunk.

Hanna helped Matthew with Brandi a lot. She would help her get ready for bed by laying out her jammies and filling the tub with bubbles. She could not bathe her because she was a foster child herself, and that was not allowed.

One evening, Hanna was sitting outside the bathroom on the floor, drawing, waiting for Brandi to take her bath. Brandi was playing with her dolls in the bathtub. "DiDi, put the washcloth down here, right here, and rub me." Brandi had her doll holding a washcloth, with the doll's hand between the other doll's legs.

Hanna looked startled. "What did you say, Brandi?" Hanna put her drawing down, stood up, and walked towards the bathtub. "Brandi, say that again, what about the washcloth?" Brandi repeated it. "DiDi rub me with the washcloth." Brandi started giggling and kept playing, making the dolls swim and jump onto the side of the tub. Hanna wondered if she was hearing this correctly.

"Brandi, what do you mean, sweetie? Who is DiDi? Does someone tell you to do this?"

Brandi nodded. "My daddy."

"Who?" Hanna asked in surprise.

"My daddy makes me wash him down there. I get candy if I put my mouth on his horsie. I get more candy, and sometimes I am allowed to ride the horsey." Brandi gave Hanna a pouty look. "It hurts sometimes to ride the horsey."

"Don't talk this way, Brandi. It's nasty talk, and I want you to stop playing that game or whatever you are doing, do you hear me?"

Brandi kept playing with her dolls and acted like it was nothing, and it was no big deal.

Hanna had no idea how to handle this, but she had to get Brandi distracted so she would move on to another subject. "Ok," said Hanna, "let's not talk about this right now." Hanna was sick to her stomach. This was too much to handle when she was not even over what happened to her in the Parris's barn. This real world that took place in foster care was way worse than living in the conditions she lived in at home with her mom. "At least I knew what to expect," she thought.

"Let's get out of the bathtub and get you to bed."

"No! I want to stay and play!"

Hanna felt herself tensing up. "Brandi, please just listen."

"I don't want to get out, you're not the boss of me!"

Hanna was scared and stressed out. "Brandi, now get out!" Brandi decided she needed to call her caseworker, but had no idea how she would do it without Matthew knowing.. "Please, Brandi."

Brandi finally got out of the tub with Hanna's help.

Hanna knew she had to maintain and think. She decided to act like nothing was wrong and tickled Brandi all the way to her room, where she had laid her jammies out on the bed for her. The jammies had unicorns on them, and when Brandi first saw them, they reminded her of her sisters. All Hanna could think of was how Lindsey and Sami would love them, but she had bigger problems than this. She had to stay focused until she found out if Brandi's behavior was true.

"What's going on in here?" said Matthew.

"Daddy, Hanna is bossing me."

Matthew looked relieved that nothing was wrong and Brandi was not giving Hanna a hard time. "Brandi, I asked Hanna to help us, and you need to show respect and do what you are asked."

"Sorry, Hanna, Brandi is spoiled, ya know."

Hanna looked at Matthew with apprehension, but approval, and gave him a slight grin. "Thank you, sir."

Matthew nodded and looked at Brandi with concern. "You can watch TV in Daddy's room tonight, Brandi, with me; that will be fun, and we will have popcorn. You can come too, Hanna, but you are older and I don't want anything said, so we can give you yours in a different bowl and you can eat in your room."

"Cool?"said Matthew. "The darn VCR doesn't work right out here in the living room with this TV, and I know Brandi has been wanting to watch the movie Frozen. I bought it for her last birthday, found it at a secondhand store."

"So you are ok just hanging out in your room with some popcorn?"

"Yes," Hanna quietly said.

"Yes! Frozen!" said Brandi, and she scurried to her room to play.

Matthew walked back into the living room and sat down, watching TV. Hanna felt very confused. Did she hear Brandi right? She needed to call her caseworker and had to think of how she could ask for the phone and not let anyone hear her. Matthew seemed so nice. He seemed like the perfect Dad. Was she imagining things because of her past with sexual abuse?

Hanna started feeling ill; her head felt like it was spinning. She felt sweaty, and she could not catch her breath. She opened her drawer and grabbed her paper dolls. She lay down on the bed and held them close to her. She hadn't had them out for a while because she'd felt safe at Tami Nickels' home. She did not feel the need to get them out until today.

Hanna's paper dolls were many, and they changed off and on. Sometimes her dolls would act out the trauma she was living through at the time. "I did put the red dress on, mama, the silky one, and I used your lipstick, just a little bit like you told me. I put on your silky black underwear and your pretty earrings."

Hanna sat and stared at the paper doll in her hands. She had blonde hair, and she had created a red dress out of her supply of paper sacks when living at home, so she put that on her, folding the tabs around the shoulders. Hanna continued sitting on her bed, admiring her paper doll, when her facial expression suddenly changed and she started acting disturbed. Her voice got deeper, and her face turned red.

"Hanna, my friend, you need to find out what is going on with your little friend Brandi."

"Nadine! Stay out of this!"

"Sneak into Daddy's room tonight and hide so you can see what is going on. You can always offer yourself to this daddy to keep Brandi from the bad things that could happen to her. You did it for your sisters, remember, Hanna?"

Hanna's facial expressions changed again, and she yelled: "No, no! Nadine!" said Hanna. "I won't do it! Stop! Stop!" Hanna began to sweat and shake, and grabbed her head, which was hurting severely. She rocked back and forth, still holding her head, whispering in distress, "Nadine, stop, no more, please."

As the pain started to subside, all of a sudden she sat up and froze, and she seemed to snap out of it. Hanna stood up, exhausted. She looked at her paper dolls all over the bed, knowing something had happened, but she did not remember anything but feeling dizzy and having a bad headache. As confused as she felt, she managed to put the paper dolls back in their bag. She walked over to her window and looked out in a trance.

One thing stood out in her mind: Nadine. She remembered creating her when she was living in the motel, and she always wanted to cause trouble, but Hanna would not ever let her fully take charge, so she fought against it. Somehow Nadine had escaped and found her way into Hanna's mind, and this time Hanna had no control to stop it.

As night fell, Hanna was terrified for Brandi, and for the consequences of the decisions she knew she had to make.

149

CHAPTER: 21

First-Day Tip-Off at Midville

The new school year was underway, and kids were swarming the halls with laughter and gossip at Midville High School. The first bell rang at the end of the day, and Sal skidded out of the classroom, down the hall, and slid around the corner. "Slow down!!" yelled Principal Scott, as Sal gave him a silly smile, high-fived him, and kept running with his smirky attitude. He was on his way to the gym for a basketball game scheduled on the first day of school. It was meant as a welcome-back event, but really it was the coaches' pre-tryout scrimmage to see who might be cut. Games officially started in October, and with the Midville Panthers winning the championship for the last five years, coaches were gearing up for more success, plus they hated the Falcons, who had agreed to play as their opponents. With the Falcons losing by three points last year in the championship, they wanted first dibs on which players would probably be cutting the Panthers, and tonight was the showdown!

Sal burst into the gym with exhilaration and barged into the locker room, running into three of his basketball buddies. "Stop it, jerk-off!" yelled Jack, flipping him off. "Sorry, I had a hard time slowing down, man, and I'm late!" Sal ran up and kissed one of his teammates, Liam, on the cheek. "I'm sorry, buddy." His future team shook their heads, rolled their eyes, and kept getting ready.

Everyone liked Sal, and he was aware of it. He dressed in a split second, slammed his locker, and jogged past his teammates, chanting under his breath, "Let's do it! Let's do it! Oh yeah, baby!" The guys were still changing into their uniforms and flipped him off one by one as Jack caught Sal's eye again. Sal stopped, smiled, and yanked Jack's trousers down. "What the hell, Sal?" Sal laughed and ran to the other side of the locker room. "It's game day, fellas, and we're gonna win this baby! Let's show the coaches what they want to see! Who's going to cut?" He was cocky, but in a great mood; he knew whichever team he played for would win. That was Sal: a winner at everything he set his mind to.

Sal ran out across the gym and slid into Coach Dan. "Seriously, Sal, not today. We didn't expect this game to be held on the first day of school. That principal Allard! He says this game is to help with school spirit, but we were thrown into utter chaos. Of course, the Falcons agreed to play us; they're still mad about the championship." Coach Dan was stressed. "We really need to get going here!" "Get going, Coach? Get going? You mean not get hammered, right?" Sal grinned. "I get it, the Falcons have already hammered us in past games, so we need to step up, right, Coach?" It was sarcasm; the last time the Falcons had beaten the Panthers was six years ago, and they still couldn't let it go. "Let's get it going, Coach!" Sal nudged him in the arm. Coach knew Sal was the best player on the team from summer workouts. A jokester, sure, but he was going to take Midville High to State; there was no doubt.

"Come on, guys, let's get the ball bouncing!" Sal was ready to get serious. The guys grabbed basketballs and started shooting. "Hey, Sal, have you seen that new girl who started yesterday in Mrs. Potter's science class?" said Hammer. "Nope," said Sal. The fun was over; it was time to focus. Hammer was Sal's best friend since elementary school, and he joked around as much as Sal. Together they were trouble, but both were highly intelligent, and most teachers and coaches knew they'd go places. They had won every game in junior high by working together, and coming to Midville High was every coach's dream.

"You have no time for the girlies this year, right, Sal?" Sal rolled his eyes. "She's a hot woman, seems shy, and I peeked down her shirt when she dropped her pencil!" Hammer laughed while he kept dribbling, trying to chat with a very focused Sal. "I want to meet her, man," said Hammer. "She seems shy, but she might be really cool." "Then go meet her!" yelled Sal, as he rose up and hit a jumper. Hammer, frustrated, waved his arms in front of Sal and swiped the ball, but he knew once the game started, Sal would find him and they'd win, and everything else would be forgotten.

Hammer decided not to waste time. He ran down the court and started shooting with the others. He glanced into the bleachers and tripped over a teammate's shoe. "Watch it, Hammer!" "Sorry, so sorry," he said. "Oh man, oh man, man, man," Hammer muttered, voice quivering. He hustled back to Sal on the far end as Sal kept getting shots up. "Sal, she's right there, holy shit, Sherlock, she's up there in the bleachers!" Sal looked up, annoyed and confused, stopped shooting, and glared. "Oh man, Sal, don't look, don't look!" Hammer stammered. "What the heck, Hammer, what is wrong with you?" He pointed with his elbow, head down. "Jesus, there's nobody there, you goofball," Sal said, then slowly glanced up to the top rows. He saw three pretty girls; two he knew, no big deal. One looked vaguely familiar, but he didn't think he knew her and moved on. "You're so stupid," he told Hammer with a grin, and took off down the court.

Since it was a last-minute game and the new band members hadn't had their first practice, Midville called in some alumni to play. The band struck up and the players took the floor.

Hanna had tried to put her concerns at Matthew Walters house behind her long enough to start high school. She barely got in on time because of the late move, but Abi pushed the paperwork through.

Hanna was impressed with Midville High; it was nicer than other schools she'd attended. A couple of girls from history class invited her to check out the gym and the game. She hadn't wanted to go, but figured she might as well try to make the best of it.

The game rolled on with Midville leading, as usual, and Sal and Hammer making a name for themselves. Hanna and her new friends headed down the bleachers without drawing a second glance from Sal. He had time only for winning. Hanna didn't notice Sal either; it had been over three years since she'd seen him in that barn at the Glasso's place. She'd been so scared and unsure that day; she'd told Sal she was a foster child, and he'd left her alone with Lindman. He'd carried the guilt for years.

Life had moved on for both of them, but that one conversation changed Sal's life. He'd heard of foster kids, but assumed they were troublemakers with bad attitudes and parents on drugs, booze, and crime. Hanna didn't seem like trouble, just scared and kind. Sal knew they could've been friends if life had been different. He'd never stopped thinking about her.

Midville won, and the team was proud but ready to go home. It was a lot to have a game on day one, especially as a surprise. "Hey, man, remember we have practice with the Trenton College band tomorrow evening?" said Hammer. "I know, for the halftime show coming up at the college?" Sal, exhausted, wiped sweat from his face and kept it short so they'd hustle, Sal's mom was picking them up. "Yes, all of us together as one big happy family. Band Day! It's kinda cool getting to see the college guys play," Hammer said. "We meet at the college gym tomorrow at 2:00 p.m. after early release. You and me, bro, we'll be playing in that gym on scholarships before we know it!"

Sal gave Hammer a love pat. "Let's go, dude, everyone in the locker room is gone." He chuckled. "I can't believe we have to head to the college before our next game. You'd think they'd let us get used to school first, but no, they toss us to the wolves. We're slammed and school just started, don't you think?" Sal slung an arm around Hammer. "Yeah, buddy, Mom's waiting. Save it for tomorrow." Hammer yawned, nodded, and followed him out. They waved Sydney down; she pulled to the curb and the boys piled in, exhausted.

Hanna thanked the girls who'd invited her. "I need to get home now, thank you. This was fun." The girls climbed back to the top of the bleachers to chat, and soon realized the gym had emptied. Hanna kept her life private; no one knew she was in foster care. She was just glad to be making friends. She waved goodbye and headed down.

"Hanna!" Lori called. Hanna climbed back a few steps. "Are you going to the college basketball game tomorrow night? The Midville band performs at halftime, and all Midville guys trying out for the team will march with the band and get to shoot a few baskets. It's in an email to your parents, make sure they're getting school emails. It's another Principal Allard motivation plan," Lori joked. "They aren't playing a full game, just marching and shooting." "I don't know," Hanna said. "I'll have to ask. I've already been out quite a while today." "My mom can pick you up," Lori said. "Come on, it'll be fun! Call me; we'll come at 5:30. The Trenton College game starts at 7:00, and we'll have a blast!"

Lori was persistent; she liked Hanna. "If you can't go to this one, Saturday night is Band Day, competitions between all the Midville high-school bands, plus a bonfire. Everything's at Trenton College. Do you have a cell phone?" "No, not yet." Lori jotted her number on a scrap of paper. "I'd like to go, but I'm not sure, I'll have to ask," Hanna said. She tucked the number into her jeans pocket. "Thanks, I'll try."

Hanna walked home with good feelings about Midville High and the friends she'd met, but she felt nervous about asking Matthew if she could go to the game and Band Day at the college. She'd already called after school to ask about staying for today's game. He'd worried, first day of school and already so much going on, but finally agreed.

Brandi sometimes slept with her dad after popcorn, but she'd be back in her room by morning. Matthew acted like it was no big deal and didn't try to hide it. Hanna never heard anything when they were together, and Brandi hadn't said a word or acted out since the day in the bathtub.

Hanna still hadn't told Abi about Brandi, and of course wouldn't tell Matthew. She pushed it to the back of her mind to see if anything else occurred and to give herself time to process. The anxiety stirred, and she felt the pull of her paper dolls.

The paper dolls weren't a secret. Back at the motel, not only did Hanna want her sisters to have dolls, she wanted a happier environment for them, so she told them the dolls were magical and that their dreams would one day come true. They were created for good. Her sisters loved the creations and looked forward to the clothes and furniture she made.

Only recently had she had an evil experience with one of the paper dolls "coming to life." She chalked it up to a bad dream as she drifted off. Hanna knew a conversation with Abi would be difficult; she feared not being believed. For now, she would keep her head down, start strong at Midville, and hope the new year stayed quiet.

CHAPTER: 22

Adopted at Last, A Sudden Cough

Lindsey and Sami were finally adopted by the Blacks, and their family felt heaven-made. Nobody would have known the children were adopted unless they said so. The girls were adopted at five and six years old. They had been in foster care for over three years but, unlike Hanna, never changed homes.

The court case was delayed a few times because of Hanna's rape. The Blacks decided not to include Hanna in their adoption plans and argued against placement with her; they didn't want Hanna living with, or influencing, their girls. They believed her trauma would affect them and demonstrated this repeatedly in court. They did agree to supervised visitation periodically after the adoption.

At one time the Blacks seriously considered asking for Hanna to live with them and adopting her as well. Then rumors in town about Hanna's rape spread. There were many write-ups in the paper about the Parrises, and the Blacks became adamant that Lindsey and Sami would be better off without the trauma Hanna might bring into their lives. They didn't want to manage the fallout from Hanna's sexual abuse and its impact on the girls. Once the final decision was made, they rarely brought Hanna up, and the girls seemed to settle into their new life without asking about her. Privately, the Blacks hoped they could stall, or stop, visitations now that the girls were legally adopted and The Department of Child Safety was no longer involved.

Canby Black, Emma's husband, mostly let Emma run the show with the two girls.

"What time is Sami's doctor's appointment, Emma?"

"Three o'clock, and we're late! Sami and I have to go. Lindsey, stay with Daddy unless your hair is combed, then you can come."

"It is brushed, Mom, look." Lindsey was getting better at doing her own hair, and Emma was proud of her, but she was in a hurry and didn't have time to help. "Lindsey, just stay with Dad so he can get your tangles out," said Emma.

"I'm no good at tangles, hon. Can you get them out?" Canby asked, looking worried.

"Lindsey, okay, fine, get in the car and hurry! Where is Sami?"

"Here I am!" Sami had red lipstick all over her lips and face.

"Sami! Where did you get that makeup all over your face?"

"Lindsey put it on me, Mommy."

"Lindsey did it? Are you sure, Sami?" Emma said, incredulous. "Canby, please grab a washcloth and wipe the lipstick off while Lindsey and I get in the car. Send Sami out when her face is clean!"

Emma started coughing as she opened the door to the garage.

"Where is that cough coming from?" said Canby.

"It's just allergies," Emma replied.

"You don't have allergies. And thanks a lot for making me deal with the girls today, you know I'm not good at it."

"Learn," Emma shot back, hurried and stressed.

Canby cleaned up Sami and sent her to the car; Sami pouted all the way, stomping her feet.

Emma ran back inside, she had forgotten her purse, and kept coughing.

"You'd better do something about that cough, Em. It sounds like you're coming down with something. We don't need that around here."

Emma grabbed her purse, backed out of the driveway, and headed for Sami's appointment. She tried deep breaths to settle down but kept coughing. "Oh great, I should've grabbed a cough drop," she thought.

Sami's feeding tube had been removed a year earlier and she was doing much better, but she still had trouble keeping weight on, so the doctor monitored her closely.

Emma and the girls hurried into the medical office. The nurse was already calling Sami as they stepped inside.

"We're here, sorry we're late," Emma said, out of breath.

"Please sign in," the nurse said, holding the door. "Sami, let's get you weighed and then head to room six."

Sami hopped on the scale. Emma and Lindsey looked at each other, relieved they'd made it on time and hopeful for good news.

Sami liked this doctor more than the others.

The exam room was cold, which made Emma's cough worse. Sami climbed onto the table as a hurried Dr. Blunz walked in, smiling.

"How is everyone? How's Miss Tami?"

"It's Sami," Sami giggled.

"You mean Carrie?"

"No! It's Sami, silly."

"I know, I know, Larry." Dr. Blunz was known for joking with Sami.

"No, it's Sami! Mommy, tell him it's Sami."

Emma coughed and laughed. "It's Sami, Dr. Blunz. You should know this!"

"Is she still eating well?" he asked.

"Yes," Emma said.

After examining Sami, Dr. Blunz looked pleased. "Mmm… I think I'll see you in about six months."

Sami stuck out her tongue with a silly grin.

"Sami! You don't do that to the doctor," Emma said, laughing, and coughing. "I'm sorry."

Dr. Blunz chuckled, then glanced at Emma with concern. "You need to get that cough checked as soon as possible."

"I know, Doctor. With kids, I keep saying I will, then something else comes up."

"Keep doing what you're doing, offer new foods and focus on healthy eating. Don't deny her much right now; we need more pounds on this kiddo."

"Yes, Doctor," Emma said, nodding.

"Goodbye, all, have a good day," Dr. Blunz said, and closed the door.

Emma, Lindsey, and Sami walked past the reception desk and out of the office.

"Wait, Mommy, I didn't get a sticker!" said Sami.

"Lindsey, please run back and get one for Sami."

"No, I want to go, Mommy!"

"Just go with Lindsey, Sami, it's fine," Emma said, impatient.

Soon both girls returned, smiling. "Look, Mommy, we both got one!" Lindsey said.

On the way to the car, Emma mentioned she needed cough drops. "Can we get M&M's, too?"

"We'll see," Emma said, buckling them in. She pulled into a drugstore parking spot. "I don't know why I keep coughing, it's driving me crazy."

Inside, a neighbor stopped to say hello.

"Hey there, woman!" Lisa called.

"Hi, Lisa, how are you?" Emma asked.

"Great. How's school going for the girls?"

"Very well. They like their classes, and I'm trying to understand the new math," Emma laughed. "Sami was just told she doesn't have to come back for six months."

"That's wonderful," Lisa said. "Congratulations, Sami!"

Sami smiled and hid behind Emma's leg.

Emma coughed again. "I'd better grab some cough medicine for this darn cough."

"I'll see you around," Lisa said. "Take care of that cough, and it's such a blessing to hear about Sami."

Emma found cough drops and medicine and queued at the register, then felt dizzy.

She set the items on the counter. "Ma'am, I'm getting very dizzy and might faint. I have my daughters with me, can you please keep an eye on them?"

The clerk looked alarmed but nodded. "Of course. Shall I call 911?"

"Yes, and call my husband. My girls know his number. Girls, go behind the counter, Mommy is sick."

Emma guided them behind the register. "Call Daddy, Mommy is sick," she told Lindsey. Then she collapsed.

The clerk shouted for her manager and redirected customers to another line.

When people saw Emma on the floor, the complaints stopped.

"If you have cash, you can check out at the liquor counter," the manager called, while clearing space for the paramedics.

"Please move back, we need room." The clerk kept the girls seated on stools and within sight.

Canby arrived as paramedics wheeled Emma toward the ambulance. She was barely coherent and sweating.

"I have the girls," he told her. "I'll drop them at Lisa's and meet you at the hospital." He kissed her forehead.

"God will carry me, Canby," Emma whispered, and closed her eyes.

"Girls, let's go," Canby said softly. "Lisa is waiting. You'll stay with her until I know what's going on with Mommy."

The girls nodded. As Canby drove, he prayed nothing serious was wrong.

Where was this cough coming from? The fainting? Emma was always so healthy. None of it made sense.

CHAPTER: 23

Is This the Right Direction?

Hanna walked in the front door after the game, and Matthew was sitting in the front room watching TV.

Brandi was at the table, playing with playdough.

"Hey, girl, how was the first day of school? It's crazy that this wasn't a scheduled game, but I get it," said Matthew.

"It was great for school spirit and gave everyone some inspiration to start the year with something positive. Brilliant for the first day of school!"

"I had a good time. Thank you for letting me go."

"So you met a few friends already, hun?"

"Yes, sir."

Matthew was trying to make conversation, but Hanna had so many doubts about him that she wasn't interested.

"Hi, Brandi," Hanna said on her way to her room.

"Look what I made, Hanna!" Brandi held up a handful of tiny clay balls. "I'm making a snowman, and a dog for the snowman."

"That's good, Brandi. Do you want me to make something with you?"

Brandi didn't answer and kept playing with the playdough.

Matthew felt content that Hanna was there to entertain his daughter and walked out to the backyard to water the trees.

"Brandi, hand me the blue playdough. I'm going to teach you to make a Smurf," Hanna giggled.

Brandi kept playing and didn't look up. She got up from the table and quickly peeked around the kitchen corner to see where Matthew had gone.

Not seeing him, she hurried back to her seat beside Hanna.

"What's going on, Brandi?" Hanna asked.

Brandi kept molding the playdough, ignoring her.

"You can tell me, Brandi. I'm your friend."

Brandi looked up; a red mark under her eye revealed a small cut.

"Brandi, where did you get that mark under your eye?" Hanna's gaze flicked to the back door to make sure Matthew wasn't coming back in.

The back door opened and Matthew walked in. "Hey, Hanna, can you watch Brandi for a little while? I have to run some errands, post office, etc.

Oh, and she has a red mark under her eye because I tossed one of her toys into the toy box and she was in the way. Poor little one, it hit her right in the face. I shouldn't have thrown it. I put ice on it."

Hanna just listened. "Sure," she said quietly.

"Okay then, girls, I'll be right back." Matthew walked out the front door and got into his car as Hanna watched from the window.

Hanna hurried back to Brandi. Up close, the mark looked worse. "Brandi, did your daddy hit you?"

Brandi lowered her head. "Yes."

"Why? What happened? Did he throw a toy at you?"

"No. He hit me."

"Why, Brandi? Why did he hit you?"

Brandi looked sad and confused. "I didn't want to take a nap with him today."

"Do you always take naps with him when I'm at school? I didn't know about this."

"Yes."

"What else, Brandi? Tell me. Your dad will be home soon. I'm here to be your friend. It's just a nap, why would he hit you?"

"Because he tried to take my pants off and I started crying."

Hanna swallowed. "Maybe he just wanted you to be more comfortable?"

"No. He tried to take my panties off too, and I told him no, and I pushed him away, and he hit me."

Brandi started to cry. "I don't want to play games with Daddy, Hanna. He hurts me when I don't want to play."

"Okay, shh. It's alright." Hanna hugged Brandi. "Listen, your dad will be back any minute. I need you to play with your playdough while I do something that will help us both."

Hanna ran to Matthew's bedroom. She had never been in there and hoped there might be a home phone. The room was a mess, with tables piled with clothes and junk. The dressers were cluttered with more items.

She scanned the room and saw several unlabeled black video cases. They looked like they could be from a library or college.

Hanna was only fourteen, but she was street-smart from years of abuse around her mother's boyfriends. Could these be videotapes of Brandi and her dad? Hanna didn't even want to think about it.

She heard Matthew's car pull into the driveway. She started to run out of the room, then grabbed one of the tapes from a stack on the dresser. She hurried to her room, slid it under her clothes, beneath her paper dolls, then ran back to the kitchen and sat with Brandi.

Matthew took his time bringing in groceries.

"Brandi, did Daddy ever make a movie of just you and him?"

Brandi nodded. "He has movies in his room of me and him. Daddy says I'm a movie star."

"Okay," said Hanna. "We're going to keep this a secret. I'm going to help you, okay? I need to get your dad's cell phone. I don't see a home phone. Can you help me with this?"

Brandi nodded.

"Good. For now, be good and keep doing your usual things, playdough, painting, okay? I'll take care of the rest." Brandi looked frightened, but she nodded and went back to the playdough.

Matthew came in the front door. "I got popsicles!"

"Sounds yummy." Hanna stood. "Matthew, do you mind if I go to a band-day event tomorrow at Trenton College? All the schools meet up and compete. I know it's last-minute, but Principal Allard is all about school spirit, like that surprise game today."

Hanna was trying to get Brandi out of the house without drawing attention and struggled to stay focused. "Midville's band will be there, and the basketball players too. I'm not sure how it all works."

"Sure, kid, sounds okay. Do you need a ride?"

"No. I met a nice girl at school and she said her mom could take me, so I could stay after school and not need a ride."

"What time would you get home?"

"Like six-ish."

"So you'd eat when you get home?"

"Yes."

"That's cool with me," said Matthew.

"Yeah, I know it's a lot. I thought I'd ask about both events, and you could tell me which one I should do. You know, one or the other?"

Matthew looked at Brandi. "I don't know, Brandi, what do you think? Should I let her do both?"

Brandi nodded with a small smile. "Yes, Daddy, both."

"Alright, then. I'm glad you're making friends. Maybe I'll go to these events with Brandi so you can have your time with your friends."

"Oh, maybe we can give you a break, and I could take Brandi with me. You could do what you want for once, right?" said Hanna.

Hanna was thinking quickly and growing nervous that things were turning the wrong way.

"Yeah, Daddy, I want to go with Hanna," Brandi added.

"Well, okay. I could use some time to get things done, maybe go visit a friend," Matthew said, relieved. "Are these good people, Hanna? You just met this girl. Can I trust them with you?"

"Why don't you have the mom call me?"

Hanna jumped in. "I have my friend's number in my pocket."

"Oh, great," said Matthew. "Let me see it and I'll call her."

"Is it okay if Brandi comes with me? You could drop her off after school and she could ride with us."

"Now that I think about it, I don't think so. That puts too much on her mom, asking her to take my girl too. Maybe another time."

"Please, Matthew. She'd have fun, and you said you could use a break."

"A break, huh? That's a long time coming," Matthew laughed. "Give me the number and I'll talk to her. What's her name?"

"I'm not sure, but her daughter is Lori." Hanna pulled the number from her pocket and handed it to him.

"Cool. I'll call her now." Matthew took the number and disappeared into his bedroom.

Hanna whispered to Brandi, "I'm trying to see if you can come with me."

"I want to go with you, Hanna. I want to see the band!"

Hanna waited for Matthew to return. He finally came down the hall grinning. "You're all set," he said.

"Great. Thank you," Hanna said.

"I didn't ask about Brandi. I don't know those people, and putting my kid on them isn't right. Another time, okay?"

Hanna felt suddenly faint, like one of her episodes was coming. "Thank you, Matthew. I'm going to my room. I have homework." She touched Brandi's shoulder as she passed. "I'm sorry, Brandi."

"Hanna, are you okay?" Matthew asked.

"Yes. I'm just tired, and I have a lot of homework."

"Well, you go, girl. We're having meatloaf tonight, some recipe I looked up. Like I have time to look up recipes, right?" Matthew laughed.

Hanna walked slowly to her room, thinking about what to do next. She lay on the bed and felt the room begin to spin.

Suddenly she sat up, got off the bed, and walked into the bathroom a few feet away. She stared at herself in the mirror, turned on the water, and cupped her hands under the warm stream. When her palms filled, she flung the water onto the mirror.

In the glass she saw, as if in the distance, her sisters calling her, motioning for her to come. "I'm trying, girls. I'm trying to get to you."

Hanna turned and walked to her dresser, unaware of what she was doing. The water was still running.

She opened the top drawer and took out her paper dolls. She sat on the bed and frantically searched for the one she wanted. She took out Nadia.

"What do you want now, Nadia?" Hanna asked, angry.

"You knew what you were supposed to do. It's your fault Brandi is getting hurt," Nadia hissed. "You were told to offer yourself. You did nothing."

"I can't," Hanna whispered, crying. "He won't take me. I'm not his blood. I'm older. He wants his child."

"You fool! You're the only one who can fix this. Offer yourself to him. Go into his room tonight and climb into his bed."

Hanna began to shake uncontrollably. A knock sounded at the door. "Hanna, dinner is almost ready! Come out, come out, wherever you are!" Matthew's forced cheer cut through her spiraling thoughts. She snapped out of it, clutching her pounding head.

"I'm coming," she called.

Hanna went back into the bathroom and noticed water all over the mirror and sink. She didn't remember what had happened, but there was no time to think, she had to get to dinner. She jumped into the shower, hoping it would help.

"Brandi, that girl is taking forever," Matthew said, impatient. "Let me call her again."

He knocked on Hanna's door, then opened it to make sure she was okay. Hearing the shower, he called out, "Girl, you better hurry, the meatloaf is getting cold!"

"I'm hurrying!"

Matthew sighed and started to close the door, then noticed the paper dolls scattered across the bed. "She's kind of old for that," he thought.

He returned to the kitchen, where Brandi was setting the table. "Brandi, look, I made homemade fries to go with the meatloaf."

"Yum, Daddy. Can we eat now?" Brandi asked, already putting ketchup on the table.

Matthew served her. "We're going to eat all this up, and Hanna will have to go get McDonald's," he joked.

Brandi smiled at her dad, then took a big bite.

Hanna knew something was wrong with her mental state. She could manage day to day, but the blackouts scared her. Still, the trauma she'd endured was nothing compared to what she might have to do to protect Brandi, and to find her sisters. The paper dolls had always been a bright spot for her and her sisters; she didn't understand why an "evil" one kept pushing her to do horrible things. She wanted to talk to someone, but how could she explain that a paper doll felt real?

Hanna joined Brandi and Matthew at the table. "I'm sorry I'm late." She smiled at Brandi.

"So," Matthew said, "I told Brandi she can't go with you this time, but I promise in the future. You need to get to know your friends and their moms a bit better. Once you, and I, feel comfortable, she can go along, okay?"

"Okay, Daddy," Brandi said.

"Dinner is really good, Matthew," said Hanna.

"Thanks, girl," said Matthew. "If you don't mind, I'd like to be excused. I'm full. Can I use the phone to sort out rides and details for tomorrow?"

"Sure," Matthew said. "Use the home phone in my bedroom, and don't look around; it's a mess." He chuckled.

Hanna stood and walked down the hall to his room. Her stomach churned. If Brandi stayed, Hanna had to decide how to get help. If she reported this, Brandi would be removed, and Hanna would likely be moved again too. It was a hard choice, but one she knew she'd have to make carefully.

CHAPTER: **24**

Is This the Girl I Knew?

"Good morning!" Principal Allard stood in front of the school, greeting everyone as they entered. "Slow down, kids! You have fifteen minutes until the bell!"

Hanna walked through the double doors of the main entryway. She liked this school. It was an older two-story building, but she felt more comfortable here than at other schools she'd attended while in foster care.

Hanna walked to school every day. Matthew house was about three blocks away, and she didn't mind the walk; it gave her time to think about everything happening at home and at school. She worried about Brandi and still felt she had to be careful about how she told her caseworker what Brandi had said. Her first class was an elective, Art. She loved painting, although lately she hadn't taken the time. Living under Matthew's roof, she rarely felt relaxed enough to do it.

The last bell rang, and Hanna took her seat in English. As she sat, her teacher, Mrs. Daniels, asked if she would take the attendance sheet to the office once it was complete. Hanna nodded. "Class, I want you to do some reading before we jump into our art project. It's always nice to know the background of the artist we'll be studying. Get out your art books and follow the directions on the board, please."

Mrs. Daniels quickly entered the missing students in her computer, sealed the sheet in an envelope, and handed it to Hanna. "I'll be so happy when I don't have to submit attendance twice, computer and paper," she sighed. "Thanks, Hanna."

Hanna smiled, took the envelope, and walked down the hall to the principal's office. There was a line, so she waited quietly, listening as a boy argued with the attendance clerk. "Look, I know I was within the fifteen minutes you allow. I'm not late every day, for Pete's sake. I have a game after school, and I need to get to class. I don't need this right now."

Hanna stepped back because the boy was turning to leave. In his rush, his backpack knocked a stack of school flyers off the counter. "Yeah, just make me late, get me kicked off the team," he muttered. As he bent to gather the papers, Hanna knelt to help and handed a few to him. He looked at her. "Thanks." He set the papers back on the counter, turned away, and gave her a second glance before walking off.

Hanna thought she might know him from somewhere. He was cute, but she'd attended a junior high in another part of the county while in foster care. Those kids would have gone to a different high school. This one, Midville High, was out of that area, but since she'd changed foster homes she now attended here.

"Where have I seen him?" she wondered. She was so worried about Brandi, and the dizzy spells kept coming, that she thought she might be imagining things. The boy disappeared down the hall. Hanna handed the envelope to the clerk and started back to class.

"I think you look really familiar," a voice said behind her. Hanna turned. It was the boy from the office. He'd come back and waited for her. "I'm thinking the same about you," Hanna said. He grinned. "Aside from thinking I've seen you before, I know you're new. I know everyone here, and I haven't seen you before. Where'd you come from? What school? Out of state or something? My parents and I travel; I've met kids while traveling."

"No, I'm not from out of state. I attended McDermott Junior High here in Midville County," Hanna said. "Yeah, there are a few junior highs within fifty miles or so," he said. "I was there for the last three years." "Where before that?" he asked. "I was homeschooled by my mom, but there were problems, so I had to move and start public junior high." He lowered his voice. "You're a foster child?" "Yes," Hanna said, glancing down. "Is your name Hanna?"

A lump rose in her throat. Was this the boy she'd met at the Parris barn, the one who'd seen how scared she was? The one Lindman hustled out? She looked into his eyes and knew. It was Sal.

"Oh my gosh, Hanna!" Sal pumped his fist. "Do you remember me? It's Sal! I've thought about you so much over the years. I can't believe this. Are you okay? Where are you living? You look good, really good!"

Hanna wasn't sure what to think, but she smiled. "I'm still in foster care, about three blocks away." "I'm sorry," Sal said. "No, it's okay. Really." She glanced at the clock. "I should get back to class." "Oh man, me too," Sal said. "I've got practice and that office slowed me down. It's so good to see you. I'll see you around, okay?" "Sure," Hanna said.

Sal jogged down the hall, turned, and waved. Hanna waved back and returned to class.

Sal slipped into his next period as Miss Welme was closing the door. "Glad you could make it, Sal," she said with a thin smile. "I would've been on time," he said, "but I went to the office within the fifteen minutes we're allowed, stood in line, and, because of that, I'm later than I would've been." "Sit down, Sal," Miss Welme said. "Oh, don't worry, I'm sitting," he muttered. He turned to drop his backpack and the open zipper spilled everything onto the floor. A few kids snickered. "Please pick it up, Sal. May I continue my class now?" He nodded. "Sure, Miss Welme. And I hope you have a good day, ma'am." She rolled her eyes and went on with the lesson while Sal gathered his things, and thought about Hanna. He couldn't believe he'd actually seen her again after all this time. Something about her pulled him in, and he intended to get to know her, and to find out what really happened in that barn.

CHAPTER: 25

Ribbons and Rumblings

"

Scoot over, Sami, and let your mom sit down! You're supposed to be up front with your team. You need to get up there now!" yelled Canby. The Black family was at Sami's gymnastics competition, and Canby was frantic because they'd arrived late. Canby had high expectations for Sami. He knew she could do anything she set her mind to, but she liked to play around and procrastinate.

Canby pointed to the front of the gym and reminded her again. "Now, Sami!" Sami had plopped into a seat, blocking her family from sitting. "I know, Daddy," Sami said. "I'm just resting before my meet." Canby pointed sternly toward the front, and Sami got up and ran to the team area, rolling her eyes in disgust at her dad. "Thank you, honey," said Emma as she sat. Emma was breathing heavily as she slid her purse under the seat. "Why are you breathing so hard?" Canby asked. "I don't know," Emma said. "Just the weather, I presume." "No, you've been like this for weeks," Canby said. Emma shrugged.

"Look!" said Lindsey. "They're lining up!" Lindsey was super proud of her sister, no jealousy at all. "Hey guys, I'm running to the restroom before this gets going, okay?" said

Canby. Emma nodded. "Dad, will you bring me back a soda and some chips?" "No," said Canby. "Waste of money. You should have brought a snack." Lindsey ignored him and kept watching the lineup. "I'll be right back." Canby made his way down the aisle and into the lobby, then stepped outside. He looked like he was watching for someone. "Hi!" said Sunday.

There she was, their neighbor, on Sunday, right outside the door, hustling in. "I'm running late, Canby. Marlee wouldn't clean her room, so I," Canby smiled and interrupted. "I figured you'd rush in at the last minute." He laughed. "Why don't you sit with us? I'm sure there's an extra seat." Canby had always felt drawn to Sunday, though he'd been happily married to Emma, until the last year. Emma could tell something was wrong in their marriage, and that Canby didn't seem attracted to her anymore. Sunday was divorced and lived with her daughter and son. Marlee darted off to her team, and Canby and Sunday walked back into the gym together, Canby deliberately leading the way.

"Emma, look who I found!" Emma smiled. "Hi, Sunday. How are you?" "Good, Emma, good. A bit late because of Marlee, but we're here!" "Please, sit," Emma said, motioning to the open chair. "Oh, it's okay, I'm going to sit closer to the front to get pictures." Emma nodded. "Okay." "Hey, Emma, I'm going to sit closer as well, if that's okay," said Canby. Emma smiled and agreed. Canby crossed the aisle and sat beside Sunday. "Finally! I can see, and get some photos," said Canby. Sunday laughed. "Do you want Emma to sit here? I can sit on the floor." "Nope, we're good," he said. He glanced at the floor where Sami and Marlee were chatting, both excited for the meet.

Canby was glad to be near Sunday. He truly did want a closer view of the competition, but he also felt a comfort with her that he'd lost with Emma. She always made him laugh. "Mama, can I get a soda from the machine?" asked Lindsey. "Where's Daddy? Oh, I see him, up front with Sunday. Can I go up there? I can see better." "Lindsey! Do you want a soda or not? Why don't you stay here with me?" "I want to sit by Daddy, Mama." "Okay." Emma pulled a dollar from her purse. Lindsey grabbed it, dropped it, then snatched it up, laughing. "Thanks, Mama." She ran to the soda machine, bought a drink, and then hurried to her dad. "Daddy!" "Hey, sugar!" "I'm going to sit here with you, okay?" Lindsey sat on the floor at his feet. "Hey, Lindsey," Sunday said. Lindsey waved.

"Daddy, look, they're getting ready for the Pledge of Allegiance." Canby loved his girls and felt incredibly grateful that he and Emma had adopted them. They had been a happy family for years, starting when the girls first came into their home as foster children. But over the last few years, Emma had changed, always tired, often unwell. Canby had urged her to see a doctor; she resisted until the fall at the drugstore. He wanted the best for her, but the ease and spark he felt with Sunday pulled him farther from Emma. He felt guilty; their family was known at church, but the attraction felt undeniable. He suspected Sunday felt it too, though she wasn't a home-wrecker. He told himself he'd speak to her privately, carefully.

Canby didn't want to hurt Emma. He respected her, but he wanted out of the marriage, feelings he'd battled for a long time. It was becoming too much. If Sunday felt the same, he would ask for a separation.

Emma sat alone for the first half of the meet. She didn't mind; she didn't feel well, and the quiet helped. Sami performed beautifully. She often placed first or second and had even taken first overall in recent meets. That made both Emma and Canby so happy, especially after all of Sami's stomach problems. Emma was proud of the progress she'd helped Sami make. She knew adopting the girls had been the right thing, even if she now lacked the energy she once had.

Emma could tell Canby was attracted to Sunday, and that Sunday felt something, too. Emma didn't have the energy to care; she needed to focus on her health.

It was time for awards. Coaches gathered to announce the winners. Bars came first: fifth place through second… and in first place, Sami Black! The crowd erupted as the coaches guided her to the top of the podium and placed the ribbon around her neck. On the floor, Sami took second. On beam and vault, she earned two more first-place ribbons, and she finished First All-Around. The gym cheered.

Everyone loved Sami and thought she was cute as a button. Many knew about her past health struggles, and they were thrilled to see her excelling. Canby and Sunday greeted her with hugs. Sami thanked them, then ran to her mom. "Mommy, look, look at all my ribbons!" "I see," Emma said. "I'm so proud of you." Emma bent to whisper, "Be thankful for your gift from God. Be proud of your friends and support them, too." "I understand, Mama. Marlee didn't place at all. She's sad. Should I give her one of my ribbons?" "That's up to you," said Emma.

Sami scampered to Marlee and sat beside her while Sunday dabbed at her daughter's tears. "Marlee, I have something for you. You're my first-place best friend." She looped a first-place medal over Marlee's neck. "Sami, that's not necessary," Sunday said gently. "Marlee is still learning. She needs to build more skills before she starts placing." "I want her to have it. She's my best friend, please." Marlee beamed. "All right," Sunday said at last. The girls clasped hands, giggling and spinning in a small circle, both of them happy.

Sunday smiled at Canby. "Sami made her night. What a special child." "Yes, she is," Canby said. "Sunday, I want to talk to you, it's important." Sunday's expression shifted. "Now?" "Yes. I'm going to send Emma and the girls home. Would you like Marlee to go with them? They're all wound up." "What do you want to talk about, Canby? You're scaring me." "It's important. I'll be right back. Come on, Sami, let's go find your mom."

Canby found Emma and Lindsey loading the car. "Emma, I'm going to stay and talk with Sunday a bit. Why don't you head home? Marlee wants to ride with you." Emma suspected what was happening, but she was tired and wanted to leave. "Okay, Canby. Please make sure the door is locked when you get home. I'm going to lie down." "It won't take long; we're discussing fundraising ideas to bring to the gymnastics meeting. I know you're tired." He walked them to the car. "I'll see you at home," he said. The girls were thrilled to have Marlee along and were still buzzing from the meet. "Goodbye, Daddy!" they called as Emma drove off.

Canby and Sunday were among the last to leave as the doors locked behind them. "What's so important, Canby?" Sunday asked.

"Well, you know, Sunday Emma has not been well lately. She has now fainted and continually has that cough, although I think it is letting up a bit.

She says it is allergies." "Well, she could be right, Canby. Allergies produce anxiety, and anxiety can do all sorts of things, including making you faint. "Yes, she will go in for some tests soon," said Canby. "That's good," said Sunday.

"What I wanted to talk to you about is you and me on Sunday. I know there is an attraction with us, and we need to talk about this." "Yes, Canby there is," said a nervous Sunday.

"I am married, Sunday, and this is not right, but I can't help it. I think we should walk away from this for a while and see how we feel after Emma finds out what is wrong.

It could be nothing, she bounces back and forth, feeling fine, and then has episodes. Maybe there is something meant to be Sunday", Canby kissed Sunday on the forehead.

Sunday threw her arms around Canby as they kissed passionately. Canby pulled away. "No, I need to go now. Let's think about this. There are a lot of people who can get hurt, including my daughters that are my entire life." "I understand," said Sunday quietly, and she turned around and walked away.

CHAPTER: 26

Night of Confessions

All the kids were arriving at the basketball game, and the band was playing loudly. It was always exciting at Midville High School's basketball games because Principal Allard demanded school spirit from everyone. The cheerleaders danced in front of the bleachers, accommodating Midville basketball fans, and finished their routine by dropping into the splits.

Hanna and Jade had been picked up by Lori's mom, with Matthew's approval. The girls were dropped off at the game and were now climbing up the bleachers to find a good seat.

"Thank you for inviting me. I haven't been to a game in quite some time." She didn't want to mention that the only game she'd ever been to had been just a few days ago. "I've been so stressed out and really needed to get out."

"Why are you stressed?" Jade asked. "I don't know. I do, but I'm not sure I can tell you. I'm not supposed to talk about certain things." "What do you mean?" said Jade. "I'll listen and won't be judgmental."

Hanna was skeptical about opening up, but she felt a sense of safety with them. "You know I'm a foster child, right, Jade?" "Yes, sorry, but Lori told me. We weren't making fun of you, Hanna. We just want to be friends, and friends share their lives and help one another."

"I'm sure it's hard to be a foster child," said Jade. "You don't even know," said Hanna. Lori scooted in to hear better. "It's starting to affect me so much that I'm blanking out at times. Yeah, please don't tell anyone, okay?" "My lips are sealed," said Lori. Jade nodded in agreement. "What do you mean, 'blanking out'?" Jade asked. "You know, kind of fainting while I'm awake. I don't know what I do during that time."

"Foster care is really a different life, you guys. Kids are taken away from bad situations. I've seen so much. But this, this is something really bad, and I'm not sure what to do," said Hanna.

Jade kept listening. "My foster dad has a little girl, and there's way too much bad stuff happening. I just can't tell you." "Hanna, stop. Look, I befriended you, and I know you've got next to zero friends here. I promise I won't say anything. You have to stop holding it in, please."

"Jade, I do know someone at this school. Well, I used to," said Hanna. "What? You know somebody here?" Jade looked shocked. "Yes," said Hanna. "I met him a few years ago when I was in foster care for the first time." "What's his name?" asked Lori. "I don't want to say." "Hanna! For God's sake!" Jade threw up her hands and then laughed. "Can we please start over here? Geez."

All the players ran out as the crowd clapped and cheered. Hanna's eyes landed on Sal. She tried to look away quickly, but she took a second glance. This time, their eyes met, and the look Sal gave her made her nervous. She cleared her throat and started bouncing her leg at a thousand miles an hour.

Hanna glanced at Sal again as he sat with his team on the bench. "Isn't he supposed to be listening to his coach? Instead, he's looking at me." "Who? Who's looking? Who are you talking about? Girl, you're crazy," Jade giggled. "Um…Sal. I mean, Sal is looking at me. And it's not me he needs to look at, but he is. Don't look at him, you guys!"

Hanna jumped up and hurried down the bleachers, leaving Jade laughing by herself. "Oh my, that girl is strange," Jade thought as she followed Hanna into the girls' bathroom. Hanna stood at the mirror, staring at herself.

"Oh, God, Jade. This cannot be happening to me. I can't do this right now." "What, Hanna? What can't you do?" "Well, like I was trying to tell you, I know Sal from a long time ago." "Sal? He's one of the cutest, most popular boys in school. You're lucky to know him. I'm…okay friends with him, I guess."

"I'm somewhat acquainted with him. He had a girlfriend last year, her name was Darcy, and she now attends another school. I'm not sure if he sees her anymore. They were, like, seriously in love." "Stop," said Hanna. "I don't need to hear that, and I probably need to go home right now." "The game just started. Come on, let's go sit down. Just don't look at him," Jade giggled.

"Okay. I want some popcorn, and then I'll go sit," said Hanna. The girls grabbed popcorn and climbed back up the bleachers. They shared the tub while Hanna nervously watched the game. Sal kept glancing her way, and it was obvious to both her and Jade. Hanna noticed that Sal was popular and that everyone loved him. He was considered a superstar basketball player.

The game ended, and the girls walked into the lobby. "Okay," said Hanna. "When is your mom coming, Jade?" "Cool it, Hanna," laughed Jade. "I want to see what happens when the love of your life walks out." "Stop, please," said Hanna. "It's just the way he looks at me. I've never been stared at like this by a boy."

"Let's get some hot cocoa," Jade said, smiling. Hanna and Jade got in line. "Hanna, you're about to find out what's going on in that wonderful, cute head of Sal's. The players are all walking this way!"

Hanna felt anxious, but she also doubted Sal would come up to her. She was a nobody at this school, and he'd been seriously in love with someone else.

"Hi, Hanna." Sal was standing behind two people in line. Hanna smiled. "Hey, that's cool you came to the game. So, you like basketball, huh?" Hanna nodded and turned toward the hot chocolate counter. She had never attended a basketball game in her life until Midville High.

"Hanna, are you going straight home? I'd love to introduce you to some of my friends." "Hey, man, don't cut in front of us! You're yelling in my ear, dude!" A couple of guys were annoyed with Sal for sliding up to talk to her. "Hey, man, thanks," Sal said, slipping right up behind Hanna anyway.

"How are you, Hanna?" Hanna looked at Sal and wanted to die. He had seen and heard things she never wanted to share with anyone. She smiled and looked down, shy. "Did you like the game?" said Sal. "Yeah," said Hanna. "Look, I don't mean to be a pest. I just wanted to catch up.

Is that cool?" "Sure, but I need to get home," said Hanna. "Where's home, Hanna? I'm sorry, I just don't know who you are, but I'd like to." "It's a few blocks away, but I really can't talk right now. I'm expected home soon."

Sal decided to let Hanna back out of the conversation. He understood, and he wanted to figure out the Hanna puzzle. "I can walk you home, if you want," he offered. "My friends are taking me, thank you."

Hanna hurried back to where Jade and Lori were standing. It took everything in her not to turn around and look at him one more time. "Can we go now, please?" Jade started laughing. "Look at you shaking, Hanna! You really like this guy." "No, please don't start rumors. Can we just leave?" Lori, still chatting, checked the time. "My mom should be out front."

Hanna, Jade, and Lori walked out to the parking lot. Sal was sitting in the back of a truck with a girl with auburn hair. Hanna dropped her gaze as they passed. Sal looked up but decided not to say anything; he didn't want to seem pushy before he'd even asked her out. Hanna had other ideas. "This is why I don't want rumors to start. Did you see? He has another girlfriend." "Hanna! Guys can be friends with girls without wanting to date them. Dork." Lori couldn't stop laughing.

They got in the car, and Lori's mom took Jade and Hanna home.

Hanna walked in the door as Matthew was rushing around the living room, picking up Brandi's toys. "Hey, Hanna, I need to run to a friend's house and help him move a washer and dryer through his back door. It's late, but he's waited a week or so, and I figured you could watch Brandi for just an hour or so. Okay?" "Sure," said Hanna. "How was the game?" he asked. "It was fun." "Who won?" "We did," said Hanna. "Great. I always like basketball. Okay, Brandi, you need to get in the bathtub and then to bed, girl." "I got this," said Hanna. "Okay, maybe an hour or so, and I'll be back," Matthew said, impatient to leave. "Brandi, you'd better be in bed by then." "I'm surprised you let her stay up this late," said Hanna. "Yeah, well, I figured I'd let her stay up and wait for you since she was upset I didn't let her go with you tonight. Let me get going. I'll be back soon. Goodnight, honey," he called to Brandi. "Bye, Daddy," said Brandi.

"I don't want to take a bath, Hanna. I already had one," said a defiant Brandi. "You did?" asked Hanna. "Yes. Daddy made me take one with him," said Brandi.

Hanna decided she needed to call her caseworker right then, but nobody was there at night, and she was at school during the day without a cell phone. She decided she would use Jade's phone the next day at school. "Come on, Brandi, let's get you in the bathtub." "I told you I had a bath, me and Daddy took one this afternoon." "Okay," said Hanna. "Get in your jammies, then, and I'll help you with the puzzle you were working on earlier this week. Then it's bed."

Brandi got dressed for bed, and they put a few pieces into the puzzle. Brandi jumped into bed, and Hanna tucked her in. Matthew wasn't home yet, and Hanna decided to question Brandi about her bath.

"Hey, little one, did you use bubble bath this afternoon?" "No," said Brandi. "We used soap, and Daddy made me put it down here." Brandi pointed to her privates.

Hanna's stomach turned, but she'd expected it. "What did you wash with?" "A washcloth," said Brandi. "Did you have a good nap? I bet you were tired." "I was, but Daddy wanted to play horsey, and it hurt, so we didn't play long. Then he let me go to sleep."

"Brandi, listen to me, little one. Please don't tell your daddy about anything we talk about, okay? Your daddy is doing very, very bad things to you. Tomorrow, when you wake up, I want you to pretend your tummy hurts. If Daddy tries to play games, tell him you feel very sick so he'll stay away, okay?" "Can he give me chicken noodle soup?" "Yes, he can give you food and drinks, but you'll be too sick for anything else. No baths, no naps, no horsey, okay? You have to be very sick so he'll stay away. I'm going to get help. Do you know what to do in the morning?" "Yes. I'll have a tummy ache," said Brandi.

"I'll be leaving for school, so I'll try to remind you, okay? Your daddy just drove up, so let's be quiet and go to sleep. I'll see you in the morning." Hanna kissed Brandi's forehead, turned off the light, and walked down the hallway as Matthew walked in.

"Hey, Hanna, check this out. More playdough and paints for Brandi!" He slid the sack across the table and laughed. "My girl is playdough-happy, and on my budget I can't get much, so I stick to coloring books, paint, and playdough, cheap stuff, but at least it's something for the turd bird." He laughed. "Well, little kids love to be creative," said Hanna with a polite smile. "Yeah, I actually love clay. Believe me, I had my artistic days when I was younger, and I couldn't get enough of it," said Matthew. "You should keep up your talents," said Hanna as she walked toward Brandi's room. "Is my girl in bed?" he asked. "She's in her room, sleeping," said Hanna.

Matthew followed her into Brandi's room. Brandi was still awake. "Hey, girl, what are you doing?" "I'm reading, Daddy." "You're supposed to be asleep." "I thought she would be by now," said Hanna. "I'm sorry." "I'm sick, Daddy."

Hanna felt a lump in her throat and thought she might pass out. "Sick? You're fine. You don't look sick," Matthew said, unconcerned. Hanna gave Brandi a warning look and shook her head slightly. Matthew didn't notice. "Just a little bit sick," Brandi said, holding up two fingers. "Maybe she is getting sick," said Hanna. "There's a flu bug going around at school." "Yeah, maybe," Matthew said. "Okay, I'm going to leave you girls alone. I hope you ate while I was gone. I'm oing to watch a movie and rest." "Okay," said Hanna. "I'll make sure she sleeps. It is late." "Great," he said, bending down to kiss Brandi. "Feel better, turd bird." He rolled his eyes and left.

Hanna whispered, "Tell him you're sick tomorrow, Brandi, not tonight." "Yes, Hanna. I promise. What did Daddy bring me?" "Don't worry about that. Just be sick in the morning. I'll try to help before I leave for school. Do you want Daddy to keep hurting you?" "No." "Okay, kiddo, good night. I'll see you in the morning." Hanna kissed her good night, and Brandi yawned, turned over, and fell asleep.

With a worried expression, Hanna went to her own room. She had a massive headache and was exhausted, but she decided to take out her paper dolls. She knew she was too old to play with them, but they calmed her and reminded her of her sisters. Hanna locked her door and spread the dolls across the bed. Her head throbbed, and she felt faint.

Suddenly, she was in the middle of a real conversation with a paper doll. "You need to tell Abi what's going on with Brandi, Hanna." "I'm not sure how to handle it," Hanna said. "You'll figure it out."

Leola was a beautiful, kind, protective paper doll Hanna had created. She helped soothe Sami when she had tough nights with her feeding tube. Hanna believed Leola watched over Lindsey and Sami and kept them safe. Hanna wasn't sure Sami remembered those days anymore, but she knew Leola had brought calm when there was pain, anxiety, and confusion.

"You need to pray," Leola said.

Hanna put Leola down and prayed for Brandi. "Please, God, help me understand what to do and give me the strength to help this child. I know I have my own problems, but I can help get Brandi out of her home because of the connections I have. Please, God, watch over Brandi while I figure this out. Amen. Thank you, Leola, for caring about us."

Hanna started to put the paper dolls away when she heard a loud voice in her head telling her to go to Matthew's room and offer herself while he was watching TV. It was Nadia. Hanna stood up, slowly opened her door, and started toward his room.

"Hey, Hanna, what do you need?" he called when he heard her in the hall. Hanna stood in the doorway for a moment and said nothing. "Hanna? Are you okay?"

A sharp pain shot through her head, and then she heard Leola's voice: "Hanna, stop. Go back to your room. Brandi needs you to fight for her, and your sisters need you to keep going for them. Don't give in to Nadia. You're stronger than this." Hanna collapsed in the doorway.

Matthew jumped out of bed and sat her upright. "Hanna! What's going on with you?" he yelled. Hanna came to, confused. "Do I need to call an ambulance?" "No, please," said Hanna. "I'm fine, just overly tired." "I think you need to see a doctor and get your blood levels checked." "Okay, I will, but I'm fine. I didn't eat while you were gone like you told me to." "Well, girl, from now on, listen." He sounded concerned.

"What did you need, Hanna? You must've come for something." "I wanted to see if I could crack the window a bit to get fresh air in my room." She had no idea why she'd come and was relieved he hadn't tried anything. She figured he wouldn't have had time anyway.

"Of course you can crack your window, crazy. Now go to bed and get some rest. If this happens again, I'm calling 911." "I understand," said Hanna. "Thank you." Hanna went back to her room without remembering laying out her paper dolls, but she remembered Leola's kindness, even if she couldn't recall why. She looked down at the bed and saw Leola among the others, but Nadia was missing. She searched quickly and couldn't find her. Irritated, she put the dolls away, thinking, I need to throw Nadia away anyway; she's ridiculously evil.

Hanna got ready for bed and started to read. She was worried about her headaches, blackouts, and memory loss.

What was causing them? She decided to check on Brandi one last time. Matthew was still watching TV.

Hanna slipped into Brandi's room and pulled the covers up so she wouldn't get cold. As she lifted the blanket, she saw Nadia tucked under Brandi's right arm. Hanna jumped back, startled. How did she get in here? Brandi was asleep. Hanna didn't remember what had happened during her headache, but she knew she would never put Nadia in Brandi's room; Nadia was too dangerous.

Angry, Hanna yanked Nadia from under Brandi's arm and tore the doll to pieces. All at once, Brandi started screaming in pain. Matthew ran into the room. "What's going on in here?" he demanded. "I'm not sure," said Hanna, and then, fast: "Maybe she's running a fever or having a bad dream." "I guess," he muttered, irritated.

Hanna hid the torn-up Nadia in her hand. "Maybe she should stay in bed tomorrow. Good Lord, what a night." "Hanna, get to bed," Matthew said. Hanna walked down the hall, hid the shredded doll in her drawer, brushed her teeth, and got into bed. She took a deep breath. "Oh, great," she thought. "Now for the other thing on my mind: Sal."

CHAPTER: 27

A Bonfire Ain't the Only Thing Burning

Matthew did not fall for Brandi's claim that she was sick, so he took her to school. Hanna tried to act normally that morning and asked if she could attend a bonfire at the school football field that night. Matthew agreed as long as she was feeling well. When Hanna arrived at school, she frantically looked for Jade to use her phone, but she couldn't find her and thought she might be sick or possibly have an appointment. She borrowed another girl's phone to call Abi at the Department of Child Safety's Office.

She called Abi her caseworker and told her everything that happened to Brandi. Abi wanted to meet with Brandi right away, so she made arrangements to go to Brandi's school immediately. Abi picked Hanna up at school and took her as well. Once at the school, Abi blamed Hanna for not contacting her sooner. Hanna tried to explain she was scared, but Abi said there was no excuse.

Abi and the police took Brandi with them to the house of Matthew Walters. They explained the accusations against him and that Brandi admitted it was true. Matthew said that Hanna and Brandi were lying. Abi told Matthew that Brandi would be coming with them, and he would need to go with the police. Matthew explained that there was no other family for Brandi to stay with. Abi told him she would go to a foster home or group home, but she would be safe and taken care of. Matthew left the house crying and screaming for Brandi. Brandi was taken to a foster home and had no emotion towards the situation at all.

"I am sorry, Hanna, but we are going to have to put you in a temporary foster group home for now. We can't have you going to the same one as Brandi, and there are no other homes available for you right now," said Abi. "I can certainly believe that after my sisters were placed in another home other than mine. I have never even seen them again. Abi, someone has to have a bed for me," said Hanna.

Hanna, you have experienced a lot in the short time you have been in foster care. We are thinking a therapeutic foster home is best for you. You are going to be placed in one this afternoon, but we are not sure they have enough room for you regularly. We will find out soon, though. What will happen to Brandi? She will go into foster care, and there will be numerous court cases to find out if Matthew will keep or lose his rights as her father. He needs to lose them, Abi! Hanna, let's get you to your foster placement and talk about this some more after you are rested and moved over.

Abi, I am supposed to go to a Bonfire tonight with my friends. There is no problem with that, with all the trauma, it is probably good for you to get out. Your new temporary placement is with Ruth at Caring Hands. She is a great lady, and once she gets to know you, she will see if you are a good fit for that home.

After Matthew was taken by the police and Brandi was taken to a new foster home, Abi drove Hanna to Caring Hands and introduced her to Ruth. Ruth introduced Hanna to her new roommate, Kiley, and showed her the room she would be sharing with Kiley. Kiley was not happy she had to share a room, but decided she would make the best of it. She hoped Hanna and her had some things in common, which in Ruth's eyes were not welcome at Caring Hands.

CHAPTER: 28

Bonfire Confessions

The bonfire was blazing as Midville High School students surrounded it, laughing and buzzing about the upcoming homecoming game. It was the talk of the town every year, and anyone who was anyone was there. Tradition said many football players would ask their date to the homecoming dance at the bonfire. Most girls had made sure they looked their best, and flirtation was in the air. Hanna wore her auburn hair pulled back in a ponytail, a blue secondhand sweatshirt, and jeans with holes in the knees. She did not care about her looks that night, and she definitely did not have homecoming or boys on her mind.

Hanna and a few friends joined a crowd of kids they knew and began talking. She wasn't in the mood to be there; she was thinking about Brandi. How could she stand at a bonfire, laugh, and pretend to have fun when Brandi was being hurt by someone she trusted and loved? What bothered Hanna most was that this was the only life Brandi knew, so the child never tried to ask for help. What went on in that house stayed quiet, and Matthew Walters made sure of it. What would all this trauma do to Brandi?

Hanna had a painful past, but she became strong by stepping up to play the role of mom. She taught herself to cook, clean, care for Sami when she was sick, and pay the bills in ways she didn't want to remember. To cope, she invented paper dolls, first for her sisters, then for herself. She knew it wasn't "normal" to play with paper dolls at her age, but the dolls were therapy. She even made clothes and furniture for them, mostly from paper sacks. Hidden in her room, she played quietly. To anyone else, it might have looked like a hobby, but the dolls felt real to her; they "knew" every secret she kept.

"Hanna! Hanna!" Sal had spotted her from a distance. He jumped and waved until he was sure she'd seen him. Hanna purposely looked away, and he knew it. "Hanna, hey. What's wrong?" He reached her as soon as he could; he'd already decided he was going to ask her out that night. He was determined to learn what really happened in the barn.

"You don't seem to be having much fun," Sal said lightly. "I know you just got here, but, you know, fun." Hanna stared at him and rolled her eyes.

"Cat got your tongue?" Sal laughed.

"I just don't feel like being here, and I don't feel like talking to anyone."

"Then why come?"

She shrugged. "I dunno. To get out of the house." She looked away, eyes glassy. She was worried about Brandi and didn't know how to tell her caseworker what was happening in the Walters' house. She was sure they'd blame her for waiting. And she didn't want to talk to Sal, especially not Sal, when she barely knew him and maybe, embarrassingly, had a tiny crush.

"Well…want to get pizza? I can be a good listener," Sal offered.

Hanna didn't know him well. He was cute, but that wasn't enough. She was worried, exhausted, and scared of what Matthew would do if he learned she had told someone. Matthew trusted her with Brandi, and he liked her.

She didn't want to go anywhere with Sal, but it still sounded better than standing by a fire answering, "What's wrong?" over and over.

"Okay. Yes. Let's go."

Sal almost tripped. The girl he'd thought about for years had just said yes. "Uh, do you want to meet after the game?"

"No. Now."

"Okay, okay." He grinned, surprised. "My truck's over there. Do you need to check in with your foster dad?"

"Do you need to check in with your dad?" she shot back, dry.

He stopped short, temper flaring. "Hey. I didn't do whatever's bothering you. I'm trying to help. I'm not going to beg you to be here with me, I've got other things I could be doing."

They headed toward the parking lot, walking side by side without meaning to. Sal pointed across the street. "The limo, m'lady, is that white truck."

They reached Sal's white pickup. He opened her door; she closed it, then opened it herself and slid as far as she could toward the passenger door. "Okay," Sal thought, "she needs space." He climbed in and pulled away from the bonfire.

Away from the glow, the road was dark. Hanna said little as Sal turned onto the dirt road that led to the lake a few miles outside town. He had no plans to try anything; he felt protective of her and, if she'd let him, wanted to take her out properly.

"Why the lake?" Hanna asked.

"I like it there. It's quiet. We can talk."

"Why not a coffee shop? Take me home. I don't want to talk."

"I get it. But I'm trying to be your friend. I don't think you have many." He immediately regretted the wording.

"You don't know how many friends I have."

"Hey, don't be defensive. You're hurting. If you want to stay that way, fine. I can't read your mind."

"I can't tell you."

Hanna began to cry, hard. Sal reached out and patted her knee, then pulled his hand back, unsure. He felt awful. He was starting to care about her, but one wrong move would push her away. "What is it?" he asked gently. "I don't need details. But if it's hurting you this much, we can tell someone who'll help."

Hanna looked into Sal's warm brown eyes and lost her resolve. "It's my foster family. There are big problems, and I should've told my caseworker sooner. I'm relieved I'm not living there now, but what that creep did," She broke off. "If I speak up, they'll put me in a group home for good. Foster parents won't want me after this."

"I thought caseworkers kept details private when placing kids," Sal said. "Isn't that the rule?"

"You don't know!" Hanna snapped. Her anxiety spiked; her hands shook; her skin went clammy.

"What happened?" Sal asked softly.

"My foster dad is sexually abusing my foster sister."

"What the," Sal caught himself. He'd never faced anything like this. "You have to tell an adult. Not me, someone who can act. My parents will know what to do."

"No! Are you kidding? I don't know your parents."

"Okay, not them. But someone. We can't let him keep doing this."

"I know." Her voice dropped. "My caseworker will be mad I waited. I was scared they'd move me to a group home, and they will."

Sal pulled her into a careful hug and kissed her forehead. "Shh. It's going to be okay."

Hanna melted into him like she'd belonged there all her life. He wiped her tears with his sleeve. Soon, she fell asleep against his shoulder. Sal felt a calm he'd never known. He wasn't just helping her; being her friend felt right. He wondered who else she had.

When she shifted, he eased a flannel shirt from the back seat and tucked it under her head. She slept on, curled toward the passenger door. Worried about the latch, he stopped, checked that it was locked, and then drove the quiet road back toward town.

CHAPTER: 29

After the Fire

"Hanna, where is the group home?" Sal asked. Hanna was still asleep. "Hanna, where is your group home? Where do you live now?"

Hanna yawned and, half-asleep, told Sal the address. He pulled up, parked, and walked around to her door. This time, she let him open it. She stepped out, turned to him, and, still drowsy, said, "Thank you for listening." She met his eyes. Part of her wanted to stay, wanted him to hold her all night, but she turned and walked up the path without looking back. Sal watched, speechless, as she went inside and the door closed. He had thought about her ever since the day in the barn, and he wasn't about to let this go. He needed to see her again, soon.

"Hanna, you're late," said Ruth when she came in. "I know. I'm sorry. I was upset and talking to a friend."

"You know the rules," Ruth said. She was strict; excuses didn't count. Hanna started up the stairs, then turned halfway. "Ruth, can I talk to you in the morning?"

"About the child?"

"I need to know if you've heard from my caseworker, Abi. How long will I be here before I'm moved to a new foster home?"

"You know you were removed from your most recent home with Matthew Walters while he's under investigation for things you're aware of. If nothing is substantiated, it's possible you could return."

"No, ma'am. Please, I don't want to go back." Hanna didn't want to explain the details about Matthews and his daughter, Brandi. Her caseworker would have told Ruth what she needed to know, and the day had already been too long.

"I understand this is a lot to carry at your age, and there are still many unknowns," Ruth said. "I'll have Abi call you in the morning to go over the case. And, Hanna, get some sleep. Chores start on time."

"Yes, ma'am. Thank you." Hanna climbed the rest of the stairs to bed.

Ruth watched her go. "Nothing but the best manners," she thought. Hanna amazed her. If Ruth had any say, Hanna would stay at the group home permanently. Beneath the hurt, Ruth saw strength and responsibility, a girl who had played the parent role far too often. Residents like her were rare, but if they avoided drugs, accepted encouragement, and kept pushing, they could build a life. Hanna had something else, too: persistence. Ruth sensed she wouldn't stop until she found a path that led her back to her sisters.

It was raining the next morning, and Hanna didn't want to get out of bed. The alarm kept ringing; she kept hitting snooze. It was Saturday, and per Ruth's posted schedule on every bedroom door, chores were due. "One more minute," she thought. Watching rain drip from the eaves outside her window, she thought about Sal, and about her sisters.

She realized she liked Sal. He listened, really listened, and made her feel heard, which almost no one did. With him, she felt oddly at home, as if they'd always known each other. That comfort scared her.

They had a beginning she wished could be forgotten. Did he know about the ranch hand? How could he? If he did, would he tell others? Would he hurt her too? The farther back she tried to look, the fuzzier it all became. Since starting at this school, neither of them had brought it up. She didn't know how much he knew, or how much people on the other side of town still talked about her. After everything with the Parrises, she'd kept to herself. Even then, she'd only spoken to Sal a handful of times, but she'd sensed he was good and willing to help.

Hanna sat up, still turning everything over in her mind. She tried to remember why Lindsey and Sami's adoptive mom hadn't wanted her in their lives. Now, older, it almost made sense. For a long time, she had prayed the woman would change her mind and take her too. No one ever came. No one ever took her to her sisters, to happiness. She was scared.

DCS knew what her mother had done, how she'd used Hanna to get drugs. Her sisters had been protected; they hadn't been used or broken the way people assumed Hanna was. She didn't think much of herself. When people looked at her, she felt like they just saw a foster kid beneath everyone else's pedestal. She suspected the girls' adoptive mother had sensed that low self-worth and decided Hanna wasn't "good enough" for their family. But no matter what had happened to her, she knew Lindsey and Sami would be all right, because she had taught them to be. She'd taught them right from wrong, bathed and fed them. She had been their mom when they didn't have one.

CHAPTER: 30

Strays & Lines Drawn

"**H**anna!" Her roommate barged through the door, rude as usual, and tossed a tiny kitten onto the bed.

"Oh my gosh, where did you get this? Cats aren't allowed in here," said Hanna.

"I got it for you," Kiley said. "I felt bad, and I thought you could use a little friend since you ain't got any." She laughed.

The kitten meowed, clearly hungry.

"I need to tell Ruth," said Hanna. "She needs to know. Maybe I can keep her."

"You are so incredibly stupid. Ruth is not going to let you keep this damn cat. You have to hide her."

"No," Hanna said, her tone serious.

Kiley snatched the kitten from the bed and squeezed until it cried out.

"Stop it!" Hanna said. "You're hurting her. You don't care, you brought me a kitten, and you don't even care because it's not good enough for you."

Kiley flung the kitten toward the door. It hit, yelped, and darted under the bed.

"Get out. Get out of here!" Hanna yelled.

She lifted the phone to call Ruth, but a large, masculine hand pressed the receiver down.

"Listen, little lady," said Pete, one of Kiley's friends. "You don't want to do that. Kiley won't like it, and when she doesn't like something, I hear about it, and then I don't like it."

He smirked. "You're cute, sexy little woman, I came looking for you."

Hanna pulled her hand away. "You need to leave. Now."

"Or what?" Pete said. "Find the kitten. I'll take it to someone who'll appreciate it."

The door swung open. Ruth stood there.

"What is going on? This room has been something else since you two became roommates."

"Hey, Ruth," Kiley said. "I was just telling Hanna she needs to get the cat she brought in out of here."

"A cat?" said Ruth. "Hanna, do you have a cat in here?"

"No, ma'am," said Hanna. "Kiley brought a kitten. It's hiding under the bed because she threw it and it hit the wall."

Ruth's eyes swept the room. "And who are these boys?"

"My friends," Kiley said. "There ain't nothing wrong with them. They're nice."

"Get them out of here. Get that cat from under the bed. Hanna, to my office. Kiley, you too."

Hanna coaxed the kitten out and carried it to Ruth's office. Kiley followed and slammed the door behind her.

"Are you slamming my door for a reason?" said Ruth. "Who brought the cat into the house?"

Both girls pointed at each other.

"Ruth, I had nothing to do with this," Hanna said. "I was lying on my bed when Kiley walked in and threw a kitten at me. She had her friends with her."

"Is that true, Kiley?"

"No," Kiley said. "Hanna came in while I was getting ready to do my chores, and she had this cat. She threw it at the door and started laughing all evil."

"Is that true, Hanna?"

"No, ma'am. She's lying, as usual."

"Who brought the boys?" Ruth asked.

Again, both pointed at each other.

Ruth pinched the bridge of her nose, as she was getting a headache. She picked up the phone and called her assistant.

"Whitney, can you come to my office?"

Whitney opened the door. "Yes, Ruth?"

"Where are the two young men who were in the girls' room?"

"Outside smoking," Whitney said.

"Of course. Please keep an eye on these two while I speak with them."

Ruth stepped outside, slipped a cigarette between her lips, and walked up to the boys. "Got a light? You look about sixteen, seventeen, is that right?"

"We don't have to tell you our age," one said.

"You're on my property," Ruth replied. "You were in my girls' bedroom unannounced. That's a problem, don't you think?"

Silence.

"Which girl are you friends with?" Ruth asked. "Don't all answer at once."

More silence.

"Here's the thing," Ruth said. "Both girls say they didn't invite you inside. That makes this a possible breaking."

"Ma'am, ma'am," one boy said quickly. "Kiley found a cat under a car. We thought it'd be funny to prank Hanna. Then Kiley wanted to keep it. We just went upstairs to drop it off. The girls started fighting, it was Hanna's fault. When she threw the cat, we left. We don't want trouble."

"Whose friends are you?"

"Kiley's."

"And the cat was for Hanna?"

"Yes, ma'am."

"Fine. Hanna keeps the cat. You do not come back unless you're registered guests. And for the record, we do not allow animals. I'm making an exception because it was a gift. Now leave."

Ruth started back upstairs and realized the boys were trailing her. She turned. "I said leave. I'll call the police."

They snickered but backed away. Ruth returned to her office. Whitney was at the copier with headphones on; she hadn't heard the commotion.

Hanna, seeing Ruth's face, started toward the door to get help.

"Not you, Hanna. Stay. I want to talk to you," Ruth said.

Hanna stepped back inside, clutching the kitten.

"Those boys," Ruth said, "I don't trust them. Be careful where you go alone."

"I can take care of myself," Hanna said. "At least I got a kitten out of this." She hugged the trembling bundle.

Ruth sighed. "Bad news, you can't keep it here."

Hanna's face fell. "I can't?"

"House rules. No animals. Do you have anyone who can keep her until we place you in another foster home? I can tell your new placement, you come with a buddy."

Hanna pressed her cheek to the kitten. "Maybe Sal could."

"Sal?"

"A friend. I've known him a little while, since my last foster home."

"Think he'll take her?"

Hanna shrugged. "Maybe. I was supposed to call him tonight before all this happened."

"You get one call a day," Ruth said. "Use it."

Hanna pulled a slip of paper from her pocket and dialed. After three rings, Sal answered.

"Hey, it's Sal."

"Hi," Hanna said softly.

"Hey, girl! I just saw you and now I get the honor of your call?" Sal sounded thrilled.

"I have a question. I don't want to ask, but I don't have anyone else."

"Ask away."

"Do you like animals?"

Sal laughed. "Hanna, just ask."

"Could you keep my kitten at your house for a little while?"

"What kitten?"

"Someone gave her to me. I can't keep her at the group home."

"Why do you have a kitten?" Sal asked, still laughing.

"Long story. Can you help? You can ask your parents and call me back, but I need to know tonight."

"Yes," Sal said, without hesitation. "When should I come?"

"Now, if you can. I need a place for her right away."

"I'll be there in thirty minutes," he said, and hung up.

Hanna grinned. "He said yes."

"Good," Ruth said. "I'll work on your placement and note that you come with a buddy."

"Why can't I just live here?" Hanna asked.

Ruth chose her words. "This place is tough. Some of the meanest, most criminal minds pass through. You deserve better. A short stop is fine, but not long-term."

"Please consider me. I'd be no trouble."

"I believe that," Ruth said. "We'll see."

Across town, Sal hung up and practically bounced. He finally had a concrete way to help. "I'll build the best scratch post," he muttered, heading for the kitchen.

"Mom!" he called.

His mother hurried in. "What?"

"Hanna called. I'm going to keep her cat until she's placed in a new home."

"Sal, your dad and I asked you not to get involved. You'll fall in love and end up in a mess."

"Mom, I really like her. She's funny. And it's just a cat."

"Your father won't allow it."

"She has almost nothing," Sal said. "This matters to her."

His father's car pulled in. Sal jogged outside.

"Dad," he said, "Hanna needs us to keep her kitten for a bit."

"Sal, we asked you to stay away."

"I want to help."

"Why does she have a kitten?"

"It was a gift. She can't keep it at the group home."

"Another reason to keep your distance," his father said.

Sal's jaw tightened. "So when someone has nothing, we turn our backs? That's not what you taught me."

His father hesitated. "We're worried about the association. You could choose a girl from a family more like ours."

"You don't know her goals," Sal said. "You're judging her."

A long breath. "We'll allow the kitten, for the kitten's sake. But I want your word you'll move on."

"I can't promise that," Sal said. "I'll be eighteen soon. I won't throw away what I've worked for, but I care about her."

He checked his watch. "I'm bringing Hanna and the kitten here so she can see where it will stay."

"Please don't bring her here," his father said.

"She needs to know her cat is safe," Sal replied. "I'll talk to Mom."

As Sal pulled away, his parents stood in the doorway, uneasy.

"We have to understand what happened that night at the barn," his father said quietly. "I've requested the police report."

His mother folded her arms. "Give the girl a chance."

Back at the group home, Hanna and her new roommate, Andi, earned extra time out for good behavior and went to the mall. When they returned, Hanna noticed the zipper on her purse, where she kept important things, was open. Inside was a necklace with the tag still on.

"What the heck, Andi?"

Andi sauntered over and plucked the necklace out. "Thanks for holding it."

"You put that in my bag?" Hanna asked.

"Relax. Nobody arrested you. We walked right out," Andi said.

"Give me the necklace. I'm taking it back."

"Nope. You're not getting me in trouble."

Hanna grabbed Andi's arm. "Take it off."

Andi shoved Hanna into the dresser. The lamp toppled and shattered. Hanna lunged, grabbed a fistful of hair, and yanked.

"Stop! I'll take it off!" Andi cried.

Hanna released her. Andi ripped off the necklace and shoved it into Hanna's hand, just as Ruth walked in.

"What is all this banging? Why is the lamp broken?" Ruth asked.

Andi's eyes watered, and she feared Ruth's reaction. "I mentioned something to Hanna and she attacked me." "Ok, what happened? You went to the jewelry counter," Ruth said slowly. "You had earned this outing, and now this argument is occurring?

Andi sniffed. "Hanna stole that necklace."

Ruth looked at Hanna. "Open your hand."

"Wait," Hanna said. "I didn't take this. Andi did; she slipped it into my bag. I was trying to get it back to return it."

Ruth exhaled. "Enough. Both of you clean this room and yourselves up. We're going to the store to return it. Whoever's responsible will face consequences."

She shut the door behind her. Hanna stared at the necklace, exhausted. Ruth suspected the truth, but she wanted the real story to come out where it began.

From Caring Hands to Open Arms

Sal pulled up to the Caring Hands group home and called the main number. Ruth answered.

""Sal… who is this?" Ruth's voice came sharp through the phone. "Sal, who are you, Sal?"

"I'm Sal," he said quickly. "I'm a friend of Hanna's. She asked me to keep her new kitten for her."

A pause. Then Ruth's tone turned firm, protective. "If you're really Hanna's friend, come to the front door and introduce yourself. I need to be sure Miss Hanna knows you and trusts you, especially if you're taking responsibility for an animal."

"Yes, ma'am," Sal said. He ended the call, slid his phone into his back pocket, and headed for the front steps.

He rang the bell. Two giggling girls cracked the door open, peeking around it like it was a game. Ruth appeared behind them and gently moved them aside.

"You must be Sal," Ruth said, studying him the way housemothers do, like they can read trouble before it speaks. "I'm Ruth. Hanna has mentioned you."

Sal nodded. "Nice to meet you."

Ruth's gaze softened, but her words stayed serious. "Hanna is very fond of that kitten. If anything happens to her, it will break Hanna's heart."

"I understand," Sal said. "I'll take real good care of her."

"Hanna's upstairs," Ruth said. "Room number three. Stay in the hallway and knock, don't barge in. She has a roommate, but after the mess tonight, I'm hoping that situation changes."

Sal didn't ask for details. He just took the stairs two at a time.

Ruth watched him go, a quiet worry settling in her chest. He's in a hurry to see my Hanna, she thought. Lord, let him be a good one.

Upstairs, Hanna sat on her bed with paper dolls spread across the blanket. Her fingers moved fast, cutting, folding, making tiny clothes that calmed the shaking inside her. She'd been upset with Andi earlier, and the dolls helped her breathe through it.

I should make a paper doll with a little basket, she thought, blinking back tears. So I can carry my kitten in it.

A soft knock came from the hallway.

"Hanna?" Sal's voice, careful. "It's me."

Hanna looked up, startled, then hurried to the door and opened it a crack. Her eyes were red, but she tried to smile.

"Sal," she whispered, relief and fear tangled together.

"I'm here," he said. "Ruth told me to knock."

Hanna nodded and pulled the door wider, then immediately looked down at the floor, like she expected to see the kitten dart out again.

"The boys... they were in here," Hanna said, words coming out tight. "They were Kiley's friends. Kiley was my old roommate."

Sal's jaw tightened. "What did they do?"

Hanna swallowed. "Ruth made her leave because she invited them in. They caused chaos... and they hurt the kitten."

Hanna dropped to her knees and reached under the bed. A small, frightened meow answered her fingers. She coaxed the kitten out, cradling her against her chest. The kitten trembled, ears flattened, eyes huge.

"Look," Hanna said, holding the kitten carefully toward Sal. "They threw her against the wall." Her voice cracked. "Feel her head, there's a bump."

Sal crouched, gentle as he touched the kitten's head. His expression hardened, but his hands stayed soft.

"I'm so sorry," he said quietly. "We'll take care of her."

Ruth's voice floated up from downstairs, reminding them like a clock. "Hanna, curfew is nine!"

"I know!" Hanna called back, then looked at Sal again, smaller now. "I don't want her to be scared."

"She's already scared," Sal said, keeping his voice calm. "But we can make her safe."

Hanna pressed her cheek to the kitten's fur, breathing her in. "I was thinking... maybe Maggie," she said, like naming her would make this real in a way that couldn't be taken.

"Maggie," Sal repeated, softer. "That fits."

He shifted in the hallway. "Hanna, you asked me to keep her at my place, right? So she can settle down, away from all this."

Hanna nodded, but her grip tightened, and something dark flickered across her face.

Sal reached over with a teasing smile and gently tickled her side, trying to lighten the moment. Hanna let out a surprised laugh despite herself.

"Hey, stop!" she said, swatting at him.

He carefully slid his hands under the kitten, trying to take her without upsetting her. Hanna resisted, half laughing, half panicked.

"Sal, no! Give her back! She's my kitten," Hanna snapped, then her voice dropped into something raw. "She's going to live at your house, but she's still mine."

Sal froze.

Hanna's eyes filled fast, like a storm breaking without warning. "Why does everyone think it's okay to take things away from me?" she whispered. "My sisters… the kitten… the things I love are always taken away."

Sal's face changed, playfulness gone, replaced by something steady and sincere. He eased the kitten back into Hanna's arms for a second, letting her feel control again.

"No," he said softly. "Not me. I'm not taking her from you. I'm keeping her safe for you, until you're safe, too."

Hanna stared at him, unsure what to do with kindness that didn't come with a hook.

He continued, slower. "I care about you, Hanna. I think… I think this could be something real between us. But I'm not going to force anything. Not feelings. Not trust."

Hanna's throat bobbed as she swallowed. "I don't know what I'm supposed to feel," she admitted. "Things have been so messed up for me."

Sal nodded like he understood. "Then don't force it," he said. "We'll go one step at a time. You'll know. Sometimes it's butterflies. Sometimes it's peace. Sometimes it's just… feeling safe with someone."

He glanced at the kitten again. "Let me take Maggie to my place. I'll set her up in my room, keep it quiet, and I'll check that bump. You can come see her as soon as you're allowed."

Hanna hesitated, then slowly, carefully, she handed the kitten over. "Promise me," she said, voice trembling. "Promise you'll bring her back."

"I promise," Sal said. "On my life."

He cradled Maggie close, shielding her from the hallway noise. "Now let's get the kitty to my house," he said gently. "She needs calm. And you" he looked back at Hanna, eyes warm, "you need to breathe. Curfew first. We'll figure the rest out together."."

CHAPTER: 31

Between Rules and Desire

Hanna and Sal pulled up to Sal's house, and the kitten purred in Hanna's lap the entire way.

"She is so beautiful, and I'm so happy you're keeping her for me. I've never had my own animal, Sal."

"I know you haven't, and you deserve it. You'll take care of her and love her like your own."

Hanna laughed. "I don't have kittens, Sal."

"I know that," Sal said, rolling his eyes.

Sal tried the front door; it was unlocked. Sal's mom stood just inside the entryway.

"Hi, guys," said Sal's mom. "I see you have a kitten there. May I see her? Him?"

"It's a girl," said Sal.

"May I see her?"

Sal's mom stepped toward Hanna and reached for the kitten.

"No, ma'am, please don't take her." Hanna's voice stayed polite but firm. "I'm not being rude; she was thrown against a wall earlier, and I'm worried she may be sore."

"A wall?"

"Mom, leave it," Sal said. "She's right."

"Okay then, she's cute. How are you, Hanna?"

"I'm good," Hanna said. She didn't remember meeting Sal's mom from the time she'd lived with the Blacks.

"Kids, I have to ask why she was thrown against a wall. That's hard to hear, especially if the kitten's staying here."

"Mom! It's fine," Sal said. "Somebody was mad. It had nothing to do with Hanna."

Hanna looked down, wishing Sal's mom would just disappear. She didn't like meeting new people, and parents were the hardest.

"Sal, do you want to show Hanna where the kitten will be staying?" his mom asked. "I was thinking the laundry room would be better for the litter box, because of the smell."

Hanna glanced at Sal and shook her head, then stared at the floor.

"Mom, the cat will stay in my room," Sal said. "Hanna wants her with me, and she'll be safe there. The smell can stay in my room too."

"The litter box belongs in the laundry room," his mom replied. "The kitten can sleep in your room, but keep the box where it won't bother everyone."

Sal looked to Hanna for approval.

"Okay," Hanna said shyly.

"Come on, let's look at my room," Sal said.

"Hey, young man, the door is open. I need to run some errands, so can I trust you?"

"Yes, Mom," Sal said, grinning.

"Okay, then I'm leaving. You two take care of things. And put the litter in the container immediately," she added, stepping outside.

"So this is it," Sal said once they reached his room. "Kind of a mess, but it'll do, right?"

Hanna thought the room was great. She could tell he loved sports from the posters and photos on the walls. His shelves were filled with trophies and ribbons of every color.

"Where will she sleep?"

"In my bed with me," Sal said with a grin. "I'll sleep on this side, and she can sleep on my pillow."

"Okay," Hanna said, smiling. "We should give her Purina Kitten Chow. I've seen that at the store and on TV."

"Let's go buy it."

"I don't have any money, and I don't want to ask you." Hanna's worried look returned. "How am I going to feed her without help?"

Hanna's eyes filled with tears. Sal pulled her into his arms and whispered in her ear, "I'm not going to let anything happen to you or this kitten."

"But your parents…" Hanna whispered.

"They understand where I'm coming from. You don't need to worry."

Sal kissed her cheek and brushed her hair back. "I fall more and more for you every day. It's like God put you in my life, and I can't think about anything else."

"I don't want to cause problems with your family," Hanna said.

"I love your heart," Sal said softly. "Trust me, I care about you."

They held each other, letting the moment settle until their breathing slowed.

Sal looked into Hanna's eyes and knew he had to make love to her, but his parents were down stairs.

He quietly shut the door, continuing to kiss Hanna. Shsh they will not come up here. Hanna melted into Sals arms as he continued to make love to her.

He didn't care if his parents walked in; he loved this girl.

The love making was beautiful, but Sal knew they both had to get moving. Sal kissed Hanna on the cheek.

Hanna nodded and set the kitten on the bed. Together they straightened the covers, sharing small, contented smiles.

"I'm glad your mom did not come up here," Hanna said. "I don't think she likes me, and this would have... complicated things."

"You're wonderful, don't let anyone tell you different," Sal said. "My parents don't know you yet. That'll change."

"Leave the door open, your mom said so," Hanna reminded him.

"Yep, she sure did," Sal chuckled, stealing another look at Hanna's smile.

Sal and Hanna left the kitten Maggie on the bed and continued down stairs, said goodbye to Sals mom, and ventured on to the pet store.

They bought kitten chow, small dishes, toys, and a scratch post Hanna thought was adorable. At the register, Hanna looked at Sal with grateful eyes. Back at the truck, she hugged him from behind.

"Wow," Sal laughed, returning the hug and adding a long, gentle kiss. "Your heart… it might be the best in the world."

"Let's get this back to the house and then get you home. It's getting late."

When they walked in with the supplies, Sal's dad met them in the living room.

"Hi, Hanna. I'm not sure you remember, but we met a couple of times when you were at Mr. and Mrs. Parris's house. How are you doing, honey?"

"I'm good, thank you," Hanna said.

"Dad, we're just dropping off the kitten's food," Sal said, and he and Hanna headed up the stairs.

"Okay, son. Your mom told me what's going on. If we can help Hanna and this kitten, I'm fine with it. I see your door is open. We haven't even met the cat."

"Thanks, Dad. We left it open because the litter box is in the laundry room."

In Sal's room, the kitten was sleeping on his pillow.

"Awww," Hanna whispered. "Look at my baby."

"Your baby is pretty cool," Sal said, kissing Hanna's knuckles. "I'm happy for you, Deli."

"What?" Hanna giggled. "Who is Deli?"

"You," Sal grinned. "Delightful, beautiful inside and out. So you're my Deli girl."

"Okay, Sal," she laughed. "I'm the Deli Girl."

"You sure are, and this deli serves cat food!" he laughed. "Let's get you back."

"It's not my home," Hanna said quietly.

"I know, Deli. It's temporary. We have to follow the rules so you can keep the kitty you love."

They headed downstairs.

"So, is the cat all set?" Sal's mom called from the entry.

"Yes, ma'am," Hanna said.

"You know she can't stay here long. We're glad to help, but as soon as you"

"Mom," Sal cut in gently. "We know. Thank you. We need to get Hanna back now."

"Where do you live, Hanna? Is it close by?"

"She lives at a group home," Sal said, already opening the front door. "We'll see you later."

Hanna turned and offered a small smile to Sal's mom, who returned a polite, tight one.

"Get in the truck, Hanna, hurry."

They climbed in, and as she closed the door, Sal slid his arm around her shoulders. "I need you here with me."

Hanna smiled. She had never really thought about boyfriends. With everything that had happened, romance felt far away, until Sal. He seemed different, genuine. She felt she could trust him.

They drove up to the group home, and Sal walked her to the door. He wrapped his arms around her, leaning in for a kiss, just as the porch light flicked on and off.

Hanna laughed. "Company," Sal said. He kissed her cheek, nudged her arm playfully, and opened the door for her.

"We need to get you a phone," he said.

"I'm not allowed to have one, but you can call me on the main line during certain hours."

Hanna waved and started inside.

Suddenly, Sal jogged back. "Hanna! Deli!"

She turned on the porch. "What, silly?"

He scooped her into a hug. "My Deli girl, if I could keep you with me every second, I would. I don't want to leave you." He kissed her forehead, then her lips.

"I know I have to go. I love you, Hanna."

"I love you too, Sal," she whispered.

Sal ran to his truck, blew her a kiss, and headed home.

CHAPTER: 32

The Night Behind the Bookcase

Hanna stepped into the group home's living room. It was unusually quiet, and none of the girls were around. She climbed the stairs toward Ruth's office; the lights were off, and the pajama lounge sat empty. The TV and the overheads were still on. She checked the bedrooms, but no one. Back downstairs, she noticed the back door standing open. No sign of anyone outside either. Maybe they had gone out for ice cream, Ruth did that sometimes, and, since Hanna didn't have a phone, she couldn't reach her. There should be a note somewhere; Ruth wouldn't just leave without letting her know. Hanna checked the kitchen bulletin board, nothing. The dining table, nothing. She ran back upstairs and flipped on the light in Ruth's office.

The bulb was dim. She crossed to the desk and pulled out the chair to reach the lamp switch, then froze. Ruth lay on the floor, unconscious.

"Ruth! What happened? Ruth!" Hanna knelt. Ruth was breathing and starting to stir. Hanna took her hand and brushed hair from her face. A swollen lump and a gash marred Ruth's temple.

"Oh my gosh, you have a terrible bump on your head. Who did this to you?"

Ruth tried to speak but couldn't. She moaned, saliva and blood at the corner of her mouth. She lay on her side, as though she had fallen after being struck on the left side of her head. The chair must have kept her from hitting the bookcase with the right side.

Hanna stood, reached across the desk for the phone, then heard a faint scrape. She looked up. The built-in bookcase shifted forward, and a narrow panel behind it swung open. Ruth had a hidden alcove behind the shelves; she used it as a safe space in emergencies. The latch must have been left ajar. A figure stepped out, Liam, one of Kiley's friends.

"Hey, girl. Surprised to see me?"

Hanna's stomach lurched, but she kept her voice steady. "What do you want?"

Liam paced, jittery, eyes unfocused. He was high. "You. We get you a cat out of the kindness of our hearts, and what do you do? You get my girl Kiley in trouble. We get kicked out. Not even allowed back. That cool?"

"You weren't allowed in our room," Hanna said softly.

"Yeah? Oh, yeah?"

The look in his eyes turned mean. Hanna grabbed for the phone. Liam lunged, driving her to the floor and pinning her. The tile was cold. Out of the corner of her eye, she saw Ruth, still down. Marijuana clung to Liam's clothes; his sweat pressed hot and heavy against her. Hanna knew what was coming. She prayed aloud, voice shaking.

"Dear Lord Almighty, please help me. Please protect me from this monster and let me live. Please, God, don't let me die. Don't let me suffer. Protect me from this evil."

"Shut up!" Liam yelled. "He ain't gonna help you!"

He yanked her long, dark-blond hair out of the way and clamped a hand over her mouth.

"Get off me," Hanna said beneath his palm.

He punched her across the face. The ceiling fan blurred overhead, dust drifting down. Heat built behind her eyes; the room tilted. She began to shake.

"Stop shaking, stop!" He hit her again. His weight kept her pinned.

"I need some of this. You gonna give it up?"

His hands were rough and calloused from construction work; when he groped her, the abrasion hurt. Hanna kicked and twisted, trying to buck him off, but her strength was fading. His left hand stayed clamped over her mouth. With his right he ripped through the waistband of her gray cotton pants, popping the button. He struck her several more times, wrestled the pants off, and flung them behind him, leaving only her lace bikini underwear.

Hanna was sweating and bleeding from her nose and mouth. Disoriented, she thought of Sal and the future they had talked about. They were falling in love. She couldn't let this thug take her life from her.

At the bottom shelf, a heavy stone bookend sat by itself. Hanna's fingers closed around it. She swung and caught Liam on the head. The bookend clattered to the floor within his reach.

"You little, I'm going to kill you!" he snarled.

Hanna bit down on his hand. He jerked back, and she screamed. With one hand, he grabbed for the bookend; with the other, he mashed her mouth again. He got hold of the stone and slammed it into her face. Blood sprayed over them both. Darkness crowded in.

Light pricked at the edges of Hanna's vision. Liam's voice sounded far away. "It's time, brothers!" Two of his buddies, Elijah and Lucas, stepped out of the hidden alcove. Hanna recognized them from school; they had been hanging around with Kiley and Liam. They were high, too. Lucas had done time in Juvenile Detention, and Elijah ran with a gang. The three of them had finally gone too far. Hanna knew they were all at least eighteen, held back in school. Elijah and Lucas laughed and bobbed to tinny music on their phones, stepping over Ruth as if she were furniture.

"How come you get to go first, man?" Elijah said. "You bloodied her up, and now we get her?"

"That's gross," Lucas muttered, still snickering.

Ruth kept still and breathed shallowly.

"I want some of that good stuff you're getting," Lucas said.

"The bitch bit me," Liam snapped. "And cracked me with a rock. You see the bump?" He slapped Hanna again. "I'm gonna kill her, Elijah."

"Why are you doing this to me?" Hanna whispered. "Why?"

Her voice was barely there. She was shaking, bleeding, and fading. She closed her eyes and prayed silently. Please,

God, don't let them kill me. Please let me live. Please forgive me for my sins. I believe in you, Jesus.

The three of them took turns raping her while she bled and trembled. Blood trickled from her mouth and ears. When it was over, Hanna lay naked on the floor, unresponsive and unrecognizable.

"Next time, Hanna, you tell us what we want to hear," Liam said.

The trio left. Ruth heard them clatter down the stairs and out the front door. She had been struck hard and knocked out earlier, but she had regained consciousness during the attack and kept still.

Ruth rolled to her side. Pain throbbed through her skull, but she forced herself to crawl to Hanna. Hanna's breathing was ragged. She needed help now. Ruth couldn't fall apart. She had to act.

"Hanna," she whispered, "it's Ruth. I'm here. We need paramedics. You're bleeding a lot."

She staggered up, grabbed the sweater from the back of her chair, and covered Hanna.

"Those low-class scumbags," Ruth said through her teeth. "I saw them. I know who they are. They will pay." She found the phone in the corner, dialed 911, and reported a break-in, a rape, and physical assaults on herself and on a resident under eighteen.

When officers arrived and asked where the other girls were, Ruth learned they had been smoking a joint in the basement. They'd assumed Ruth was on a conference call in her office. When Ruth finished the call and couldn't find anyone, and smelled marijuana, she went downstairs and caught the girls with the three boys, who later attacked her and Hanna. When she confronted them, the boys shoved her; she retreated upstairs to call the police. They followed, and one of them struck her in the head, knocking her out. She came to later when she heard Hanna's voice and kept still, listening.

Paramedics loaded Hanna onto a stretcher. She never regained consciousness, but she was alive. Ruth rode in a second ambulance to the hospital. Once she was settled in a room, she gave the police names associated with the assault, including Hanna's roommate, Kiley. Ruth would not let them walk free.

Ruth had been through a lot. The assault dredged up memories of her own past; her grandfather had raped her; the court case was grueling, but he was convicted before his death. She had channeled that pain into purpose: working with foster children and directing a group home. Seeing Hanna brutalized left her furious and fiercely protective, as well as deeply sympathetic.

Ruth loved Hanna. She saw a smart, strong young woman with a future, and she knew Hanna carried a heavy ache about her sisters, separated into different homes. Ruth had been trying to arrange visits, but the adoptive family hadn't returned her calls. She hoped to bring Hanna into her home permanently.

The hospital wasn't busy. Hanna was admitted to intensive care immediately. She did not respond to stimuli. The team moved quickly: a concussion, a broken collarbone, multiple facial lacerations, and a broken nose. Imaging showed internal injuries. Tests also revealed that Hanna was pregnant, and, miraculously, the baby appeared unharmed.

While Ruth rested, officers told her they intended to pick up the suspects that evening. Paramedics asked for a family contact. "She's a foster child," Ruth said. "She only has me and her caseworker. She's between homes right now. I'll give you the caseworker's name, but for now, it's me."

The medical staff did everything they could to stabilize Hanna. No one expected her to leave anytime soon.

Across town, Sal sprinted up the steps of his house, eager to call Hanna before lights-out.

"Son, don't slam the door," his mother called.

"Sorry, Mom!"

He dialed the group home. No answer. Odd for that hour.

Six months had passed. Hanna was now in a rehabilitation center and had not spoken a word since the attack. Her face had required full reconstruction.

Ruth recovered and made certain all three assailants were prosecuted. They were eighteen and nineteen.

Sal learned what had happened. For six months, he went to the hospital every day, but he was never allowed to see Hanna or receive updates; he wasn't family. He waited in lobbies, sat in the chapel, and prayed. He never found out she was pregnant. Eventually, his parents wore him down: she was a foster kid, a bad choice, time to move on. He finally stopped going. He told himself she would never get better; too much time had passed.

He decided to pursue his original plan: college for aerospace engineering. He'd earned a basketball scholarship, and with his parents' support, he had a free ride. Hanna had missed the end of the school year, graduation, and the ceremonies where Sal received his awards and scholarships. He had never intended to leave the state, or leave Hanna, but now he needed distance to dull the pain. He began to pack, sadder than he had ever been in his life.

CHAPTER: 33

A House for Hope

Riley chuckled. "Healthy as a horse, that woman. Three times a week, she goes to yoga, and she walks daily with her friend Mindy. I get tired of the healthy cooking, though! Let's go out tonight and get steak and potatoes, and that garlic bread you like so much!"

"Sounds good," Sal said. "Text Mom and let her know we'll be there in a few minutes."

Sal texted his mother, and as they drove up the circular dirt driveway, she was waiting on the porch swing.

"Hey, hey, hey! It's my boy!" Sydney called, hurrying down the steps. She threw her arms around Sal.

"Hi, Mom," Sal said, kissing her cheek. "I guess Dad wants to go out to eat."

"Yep, and that sounds good to me. Let me shower and get dressed."

"Oh, no," Riley said. "That'll take two more hours. We're hungry now, hon."

"Okay, okay, then I'll just change my earrings and I'll be ready."

"Okay," Sal laughed. "Dad, let me take my luggage in. I don't have much."

"Okay, son, let's hurry and get some grub!"

Inside, Sal headed upstairs to his room. Maggie, the cat, greeted him at the door. She was five now and a great cat. He had thought about rehoming her when Hanna was hospitalized, but he couldn't do it. Now he and his parents were attached to her.

Maggie hopped onto the bed to greet him. Sal stroked her and rubbed her belly while she rolled and purred, thrilled he was home. He was thankful his parents had kept her while he was away at school; they loved the little cat, too.

Sal put his things away and hung some clothes in the closet. He tossed his backpack onto the bed, and Maggie immediately climbed on top of it, purring.

"Get off my backpack, Maggie."

Maggie lay on the backpack and refused to move.

"I said, get off." He nudged the stubborn cat so she'd step aside. Maggie slid off, turned away as if offended, then climbed right back on.

"What, girl? Why do you want this backpack?"

He sat on the bed. He was hungry and stressed and, for a moment, thought about staying home. Then he remembered the newspaper in the front pocket. He unzipped the front pocket and pulled it out.

Maggie hopped off the backpack and kept purring, refusing to stay off his lap as he scanned the newspaper.

"Girl, what are you trying to tell me? Is this Lindsey? Is this Lindsey, Hanna's sister?"

He studied the photo, absently stroking Maggie. "Look, girl, she does look like Hanna, huh? If it is her, I wonder if she even knows what happened to Hanna. I bet she doesn't. If this is her, she's a famous violinist now."

"Maybe I should try to contact her and let her know about Hanna. It's the best thing I can do for Hanna. If Hanna were coherent, she'd do it herself. What do you say, girl? This would make us both feel better, right?"

He slid the newspaper back into the front pocket of his backpack.

"Come on, you two!" Riley yelled. "I'm getting cranky!"

"Okay, here I am!" Sydney laughed, stepping out of her room with new earrings on.

"I like your earrings, Mom."

"Thank you, honey, let's go eat."

They walked out to the car. "Does Mexican sound good?" Sydney asked.

"No, steak and garlic bread, Mom. Dad's request!"

They all laughed and headed out to find a good steakhouse.

CHAPTER: 34

Threads of Faith and Practice

"**I** am getting so tired of these buses," Lindsey said, slumping against the window. She was exhausted from traveling, and they still had one more concert to perform.

"Mom, do we have to keep driving? Can't we stop at the next hotel to eat and rest?"

"We can't," Emma said. "We have to get to the next city so we're on time tomorrow, your performance is at 3 p.m. You know how packed our days are; it'll be here before you know it."

"All right, all right. If I can finally get phone service, I need to call Sami and see how Macie is doing. Macie is not easy, Mom."

"I know," Emma said with a small grin. "It won't be perfect for a while. She has to get used to our ways. Remember, she was almost adopted, she was very used to her last home, and loved those foster parents. We have to give her time. There should be an adoption date set soon. It makes me feel good that you love that little girl so much. I feel like we did another great thing, taking her out of the system."

"Mom, I've never seen such a beautiful fit beyond blood."

"You're adopted, silly," Emma said, puzzled and fond.

"I know," Lindsey said, "but it doesn't feel that way. When you see it in another child's eyes, it looks different."

"Different? What do you mean?" Emma asked.

"We're watching Macie become part of our family right in front of our eyes. It feels meant to be. She loves us so much already, Mom. It's like we've been together forever."

"I agree," Emma said, "but remember there's a honeymoon period. She'll be on her best behavior while she adjusts. It may not always be like this; her true feelings will come out."

"I know, Mom. I'm just saying she already feels like family, almost like she's blood. Like one of my sisters, related to me, Hanna, and Sami. It's weird."

Lindsey was tired, and Emma saw emotion welling. "That's how I've always thought of you and Sami," Emma said softly. She laced her fingers over Lindsey's hands. "God brought you into my life for a reason; we were meant to be a family. Macie is part of God's plan. She's supposed to be with us."

"I know, Mom, but I still don't know where Hanna is. It's been too long." Tears slid down Lindsey's face. Emma felt a sharp pang of guilt for keeping a distance because of Hanna's trauma.

"Your dad and I have talked," Emma said, "and now that he has more time because of his work schedule, he says we need to find out where Hanna is. I pray you're not blaming me, Lindsey. I loved you both so much. Hanna had such a past and had been in several foster homes; I couldn't bear the risk of it affecting you."

"She's my sister, and I don't blame you, Mom, but you should have at least let us keep in touch as we grew up. Courts allow siblings to do that." Lindsey wiped her cheeks. "The court could have worked it out if you'd agreed."

"Lindsey, I couldn't," Emma said, tears gathering. "You both were my world. I couldn't take another chance that anything would scar you. I'm sorry. I hope you forgive me."

"Mom, of course. I love you. You don't have to ask for forgiveness," Lindsey said. "When you love someone, you communicate and work things out. We'll find her, on our own if we have to."

"I know you will, honey. Now that you're older and understand so much more, I have no trouble helping you find her. You were so little when you were adopted, and your father and I would not let anything hurt you. It was all about loving you."

"I know, Mom." Lindsey smiled as the bus slowed to a stop.

"Ugh, we're stopping. I need to stretch my legs," she yawned, wiping away the last of her tears.

"Okay, let's both get out and grab coffee and hot chocolate at that restaurant over there," Emma said.

"Well, tell the driver, what's his name?"

"Lewis," Lindsey said.

"Lewis, we're getting coffee and hot chocolate," Emma called. "We'll be right back." She moved the canvas bags from the aisle, signaled Lindsey, and hopped off the bus.

"Okay, ma'am," Lewis said.

Lindsey and Emma hurried to the coffee shop. Emma grabbed a newspaper from the plastic rack by the door. "I want to see if we're in here; we're supposed to be," she said.

"Great, let's sit," Lindsey sighed.

They slid into a comfortable booth. Lindsey opened the newspaper to the entertainment section. "Yep. Look, Mom, here we are."

"Fantastic," Emma laughed. "I look great." She read their names aloud. Lindsey grinned. "Mom, I never thought I'd end up a violinist, that's for sure." Emma smiled and winked.

They sipped coffee and hot chocolate, slipped off their shoes under the table, relaxed and giggled, and waited for the driver to fetch them.

"All aboard!" Lewis called from the doorway, motioning for them.

"Okay, Mom, we can do this, right?" Lindsey said, stifling another yawn.

"Yes, we can, my dear." Emma was exhausted, too, but she didn't want Lindsey to see how easy it would be to procrastinate and lose momentum. Lindsey was at the beginning of her career, when everything mattered. She was the best, so young, talented, and unique, and everyone knew it.

On the walk back, Emma's thoughts turned to the past. Lindsey would be more at peace and more focused on her future if she had answers about Hanna. When the girls were little, Emma had simply said no: "We're not bringing Hanna into our lives, not after what she endured as a young child."

What Hanna saw, what she suffered, and what she became in survival mode were not good for Emma's babies. Emma had hoped that bringing Macie into Sami and Lindsey's lives would help keep their minds off Hanna.

The girls had been so little when they entered foster care. Emma believed they didn't remember much; she wanted it that way. But Sami and Lindsey remembered more than Emma knew. They remembered Hanna mothering them, cleaning scraped knees, feeding them, soothing them. They remembered their mother sleeping most of the time with her boyfriends and not getting up until night again. They remembered Hanna washing dishes with a bar of soap and making a bed for the three of them in the closet to stay warm because the motel room was so cold. They remembered the paper dolls and how Hanna used them to help the girls through hard moments. These were the good things about their big sister.

They also remembered Hanna being carried out in the middle of the night by their mom's boyfriends. They didn't know what was happening; Hanna would slip back in later and crawl into bed. They had been so little, but they knew Hanna protected them. She was their mommy when their mother was not around.

Back on the bus, Lindsey buckled in. "I still need to call Sami and check on Macie, and tell her the ideas I have for finding Hanna," she said.

"Go ahead, honey," Emma said. "We need a game plan to start looking seriously."

"It's been years, Lindsey, be careful," Emma added gently. "Hanna will be grown now, and we don't know what her life looks like."

"Or she could have been adopted by a great family, like Sami and me," Lindsey said. "I don't understand why she didn't look for us. She would have. She doesn't give up. None of it makes sense. I hope nothing happened to her." Concern creased her brow.

"I don't know," Emma said. Adoptions are private. She knew you had Dad and me, remember your biological mom's funeral? Even then… Maybe she was just too young to figure things out, Lindsey."

"Please don't think I kept you away from her," Emma added quickly. "Well, I did, by not contacting her myself. But she never tried to contact your father or me either. DCS is so secretive, Lindsey." Emma swallowed. She had done everything she could to keep Hanna away. Now she felt she would face the consequences. I deserve this, she thought. Maybe Macie will help us unite as one, and maybe we'll find what happened to Hanna.

"We're good people, Mom," Lindsey said.

"I know this. Stop," Lindsey shouted gently through a tearful laugh. "Just get a hold of Sami, and we'll make the plan."

Lindsey squeezed Emma's hand with assurance. Emma, oddly lighter, felt a cautious hope; this might be the beginning of a new chapter for all of them.

CHAPTER: 35

Paper Dolls, Quiet Scars

Hanna had been discharged from the psychiatric hospital after five years of treatment. She looked like herself again, healthy and bright, after two cosmetic surgeries. Ruth picked her up and brought her back to the group home. Hanna was twenty-two now, and she had nowhere else to go. Ruth had stayed by her side through the trauma, the assault, the decision to place the baby for adoption, and the long, slow recovery. Hanna still struggled to live in the present; what had happened five years earlier felt as if it were happening now. Everyone agreed to go slowly so she could make sense of things again and catch up to today's world.

Hanna continued to improve. She began walking again; she had once been a jogger. She went to church on Sundays and had lunch with old friends. Most of the people who visited were from the group home years ago; when they learned she was back in town, they came to see her. A few from church and from her old high school stopped by as well. Her friend Lori even visited with her two children. Everybody loved Hanna, and everyone was told to act as if they were meeting her for the first time so she wouldn't be stressed or triggered by memories. The psychiatrist said this approach was temporary.

Hanna still played with her paper dolls, and now she mentioned them openly to the girls in the house. "They're my best friends," she would say, and in many ways, they had been. Ruth told the girls to go along with it and listen. They didn't have to talk to the dolls; they only needed to respect Hanna.

Hanna had different names for the dolls, and she had recently made one that was pregnant. Ruth wasn't sure whether these dolls were the old ones from the past or new ones Hanna had created. She mentioned it to Hanna's psychiatrist, who thought it might lead to a breakthrough. Hanna never showed distress about placing her baby for adoption; because the pregnancy followed the assault, it was as if her mind had protected itself by keeping it out of reach. Hanna had been in such a fragile state when the baby was born that she barely remembered any of it.

There were nine dolls now woven into Hanna's life: her sisters Sami and Lindsey; Leola, who had always stood by Hanna; Mommy; a cop; a teacher; a doctor; a farm boy; and now a baby. When Hanna introduced her handmade friends to the roommates, she called Lindsey and Sami her sisters, and she referred to her mother as Mommy. The cop, the construction worker, and the farm boy were introduced too, but without the warmth reserved for her sisters, Leola, and Mommy. Hanna never brought out the baby; it stayed sealed in a plastic bag.

Few people knew about Hanna's dolls from the past, only her sisters, and a handful of others who had glimpsed them at moments during Hanna's trauma. When those glimpses happened, people wondered but never had an explanation. Ruth had heard from a therapist that there had been dolls, but she hadn't truly encountered them until now.

Hanna was settling back in when Ruth ran into a bookkeeping snag. Hanna was great with math, so Ruth called her upstairs. It was the same office where both Ruth and Hanna had been attacked. The psychiatrist hadn't said whether the room itself might trigger Hanna. Ruth needed her help and, in the moment, didn't focus on what the setting might do to her psychologically.

"What do you need, Ruth?" Hanna said. "Ugh, I was just running and my legs are about to give out! Then you call me up here and I climb more stairs." She laughed.

"Hanna, I need help with a math problem. You know I rely on you for the tricky accounting. I'm not sure where you picked it up, but you definitely have the skill."

"I remember taking accounting in high school. What do you have?"

"Come over here behind my desk so you can see," Ruth said.

"Okay," Hanna answered, walking toward the desk.

All at once, Hanna froze. Her hands fidgeted as she rifled through her backpack, the one she wore for runs, with a water bottle clipped to it and a front pocket for small things.

"Hanna, what are you looking for?" Ruth asked.

Hanna didn't answer. She plunged her hand into the front pocket and pulled out a small bag of paper dolls. She held up the cop. When she spoke again, her voice was low and different.

"Why do you need to talk to Ruth?" the cop asked.

Ruth blinked. "I don't need a cop. I just needed to ask Hanna a question." She tried to play along; she could see Hanna sliding into another state.

"Ma'am, step aside," Hanna ordered in the cop's voice, lifting the doll in front of Ruth's face.

She pulled another doll, a woman, and gave it a firm, protective tone. "Officer, this is Ruth, and she's kind. I don't want any trouble," Mommy said.

"Who are you?" the cop demanded.

"I'm Mommy, and you leave my girl Hanna alone."

"Ma'am, I was called because there's a problem with this Ruth asking Hanna to do something she doesn't want to do."

"Officer, I'm Mommy, and she'll do it anyway. She'll do it right now," Mommy snapped.

Hanna lifted the Mommy doll close to her face and raised her voice. "Hanna, you will comply with this officer," Mommy barked. "Take off your clothes now and get on the desk."

Ruth stepped forward gently. "Hanna, Hanna. That's enough."

But Hanna stared past her and pulled another doll from the bag.

"Mommy, no. Hanna, no!" a small, urgent voice cried. It was Lindsey. "Get Sami and put her in the closet, someone's coming!"

Hanna grabbed a second girl doll. She dropped the cop on the floor and hurried the two girl dolls to the office closet. She opened the door, stepped inside, and pulled it shut.

Ruth stood motionless. The office was a trigger; she could see it now. Was this good or bad? The scene unfolding wasn't about the recent assault; it was older. Mommy. A cop. Her sisters. Something from home.

The door creaked. Hanna emerged from the closet. "Ruth, it's fine. I have the girls hidden. Nothing will happen to them," she said, steadying herself. She pulled out another doll. "This is Miss Danny, she's a teacher at my school. I don't always get to go because Mommy doesn't get up to take me, but I love school."

She produced another. "This is Mommy's boyfriend, Austin. He's a construction worker. He's looking for Mommy. Have you seen her, Ruth?"

Ruth didn't want to say Mommy had fallen from the desk in this very room. "No. I don't know where Mommy is, Hanna."

"Where is Hanna?" Austin rumbled in a low voice.

Hanna dropped to her hands and knees and crawled under the desk toward Ruth. "Don't let him see me or my sisters. Please, I'm begging you." She kept low, scuttled back to the closet, and slipped inside again.

Ruth waited, heart pounding, unsure what to do or say.

A soft knock sounded from inside. The door opened, and Hanna crawled out. She stood, cradling the two sister dolls, kissed them both, and whispered, "Marvin came and saved us. Marvin the farm boy. I like him. I hope he likes me. He plays basketball. I want to get married and have babies with him, Ruth. Shh. I think we're safe."

Hanna drew a breath and looked around. "Ruth, what did you need again? Why are my paper dolls all over in here?" She glanced at the desk and the floor. "Were you looking at them? They were in the front pocket of my water backpack."

"I didn't touch them," Ruth said.

"Those girls in this house, they'd better leave them alone," Hanna muttered, then smiled. "Maybe I should make some for them." She stepped behind the desk. The episode passed; Hanna's voice and manner returned to baseline.

"Move over and let me give you the answers to your accounting problem," she giggled.

Ruth smiled and pulled up a chair beside her. She could hardly wait to call Dr. Richmond and describe what had just happened. Confusing as it was, it looked and felt like a breakthrough.

CHAPTER: 36

Overture to an Ovation

al had finished dinner with his parents. All they wanted to talk about was school, girls, and his college experiences. They kept bringing up companies that offered aerospace engineering apprenticeships. Sal wanted to go home; he felt his life was his business, not theirs. He tried to be courteous, staying engaged with their questions, but irritation crept in. They had always been great parents, but as he had gotten older, he no longer felt the need to report every minute of his day.

They often brought up Hanna when they were together. Strangely, this time they had several chances to mention her, and didn't. Sal, however, couldn't stop thinking about her because of the newspaper article tucked into the front pocket of his backpack. He was eager to get home, reread it, and figure out how to contact Lindsey and her parents. Lindsey and her parents.

"Well, that was a great, if fattening, dinner," Sydney said.

"You're absolutely right, Mom. I'm stuffed. How about you, Dad?"

"I'm stuffed too," his dad said. "What do you say, ice cream down at the coffee shop?"

"Dad, I just got in today, and I'm really tired."

"You don't get tired," his dad laughed.

"I do tonight. I'm sorry. Why don't you two go get some ice cream and just drop me at home?"

"Sure," Sydney said.

They pulled up to his parents' house, and Sal hopped out. His parents headed for ice cream while he unlocked the front door. Inside, he kicked off his shoes and ran upstairs to his room. He lifted his backpack off the hook on the back of the door, tossed it on the bed, and unzipped the front pocket.

Maggie was there in an instant, purring and rubbing against the backpack and Sal's arm.

"I see you, I see you," he laughed. "I'm getting it out."

Sal slid out the folded newspaper and began to read. The caption and the article are about Lindsey Flemming. So she wasn't using her adoptive last name, Black. "This is her, girl, it's her," he told Maggie. The piece also mentioned she lived in Midville County.

He tried directory assistance for Emma and Canby Black, unlisted. He tried Google; the names surfaced with mismatched numbers and ages. A few Canby Blacks popped up in nearby cities. He called them all. No luck.

"Maggie, I think we've got another way to find Lindsey." One listing showed a Canby Black at Garth and Company, near Midville High School. Sal had driven by it; it was close. He'd go there in the morning.

He woke early, determined to stop by Garth and Company and ask for Lindsey's father. He parked in the garage and walked to the receptionist's desk.

"I'm here to see Mr. Canby Black. I don't have an appointment, but this is important."

"Mr. Black isn't in right now; he's out to lunch," the receptionist said.

"Oh, I see, ma'am. With his daughter, Lindsey, I'm guessing?" Sal smirked and winked.

The receptionist looked puzzled, then nodded. "I'll let him know you're waiting so we don't schedule over him."

"Thank you. I'll wait," Sal said.

He settled into a yellow leather chair and picked up Sports Illustrated. He wasn't leaving until he spoke with Canby Black. He had to know whether this was the right Lindsey, and whether she knew anything about Hanna's condition.

About thirty minutes later, the receptionist called him to the desk. "Sir, Sal Mackhe? Mr. Black will see you now."

"I thought he was at lunch," Sal said.

"He was," she replied. "He came in the back way, and I didn't see him. He'll see you now since he's free."

Sal followed her through the offices.

"Hello, Sal, is it?" Canby Black said, offering his hand.

"Yes, sir." They shook. "Thank you for seeing me."

"You didn't make an appointment," Canby said. "Julie said you seemed eager. How can I help?"

"Mr. Black, may I call you Canby?"

"Go ahead."

"This may sound like nonsense, but I need to know if your daughter is related to my former girlfriend."

"I'm very busy, Sal," Canby said, though his tone was not unkind. "I'm not in the habit of discussing my daughter with strangers."

"Yes, sir. Let me explain. I'm looking for a Lindsey who has a sister named Hanna. The Lindsey I saw in the paper is a phenomenal violinist. When I saw her picture in The Roundup, she looked so much like Hanna, and she's named Lindsey, that I had to find out. Hanna… she's been in a psychiatric hospital, and the doctors don't think she'll fully recover. I'm visiting my parents and found the article at a coffee shop. I couldn't ignore it. Maybe God is telling me to do something."

Canby held his gaze, weighing him. "Tell me more, Sal."

"Hanna and her sisters, if I'm not spilling something you don't know, were removed by the state. Lindsey and Sami, the younger girls, were adopted by what we heard was a wonderful family. I believe you're part of that family, sir. Hanna wasn't as fortunate. She bounced through foster homes and then a group home, where she suffered terrible trauma, fell into a coma, and now lives with severe mental illness. I stayed by her as long as I could, but she didn't get better. I went back to college and tried to move on. But when I saw the article, I thought: if this is your Lindsey, maybe I can help connect them. Hanna would want that."

"Ordinarily," Canby said, "I'd ask security to escort you out for walking in here with a story like this." He paused. "But your details line up. I am Lindsey's father. I've heard the outlines of the girls' story. Two of them are ours, fourteen and fifteen now."

"Would it be possible for me to speak with your wife and with Lindsey directly, when she's back from traveling?" Sal asked.

"Yes," Canby said. "We can arrange it when she returns. Leave your number."

"Please don't lose it," Sal said, handing over his card. "This means everything to me. I know Lindsey and Sami would want closure for their sister. I still love Hanna. It's been five years, and I'm still struggling to move on."

"I understand," Canby said. "I'm sure Lindsey will want to see you, too."

Sal walked out of the office with a heavy kind of hope. The sadness never really left, but this felt like something he could do, for Hanna, for her sisters, and for himself.

CHAPTER: 37

Where's Hanna?

Lindsey's violin concerto was, as always, astonishing. Newspapers would soon be posting headlines like, "Lindsey Flemming, the Inspired Violinist, Enters the Billboard Classical Chart at No. 1."

Exhausted, Lindsey and Emma decided to order pizza to their room. Lindsey was soaking in the tub when her mom called in, "You've got a call from Dad. Please hop out and call him back; he sounds like he misses you."

Lindsey stepped out of the tub, blew out the candles, wrapped herself in a towel, grabbed her phone from her purse, and dialed her father. The call rang three times before Canby answered.

"Hi, honey. Do you have a few minutes?"

"Of course."

"I'm sure your concert went beautifully, like always?"

"Well, you know me," Lindsey laughed.

"I had someone come into the office today, his name is Sal. Now, hear me out; it may sound odd, but listen, okay? Hear me out; it may sound odd, but listen, okay?"

"Sure, go ahead. I'm listening."

"He was looking for you."

"What? Why?"

"He saw your newspaper article, your first and last name, and thought you might be related to someone. He said you looked like her."

"Who?"

"Apparently, he's Hanna's boyfriend. Or her ex-boyfriend."

"Oh, here we go," Lindsey laughed. "Please don't tell me this is some weirdo, Daddy."

"Honey, I don't think so. He said Hanna is in a psychiatric hospital and not doing well. She wasn't as lucky as you and Sami. She was in several foster homes and finally a group home, where something really bad happened and left her in this condition."

"Oh my gosh! Is he telling the truth, Daddy?"

"He certainly seemed to be. He wants to meet with you. Of course, I'll try to be there as well. What do you think?"

"Do you have his number?"

"Yes."

"Please give it to me. If this is true, I need to find out about Hanna."

Canby read the number to Lindsey. "Do you want to wait until you get home so I can be there when you call him?"

"No, I want to do it now. I'll be fine."

"I love you, honey."

"Dad, this might be the last piece I need for closure about my unstable early life. Jesus, I barely remember it, but

I do remember Hanna taking care of me. We played with paper dolls all the time. I was little and tore them up, she got so mad," Lindsey said, eyes filling. "I think she made them out of paper sacks."

"Good night, honey," Canby said.

"Thank you for telling me," Lindsey said and hung up.

Lindsey was bone-tired and didn't want one more thing on her plate, but this was probably the most important call she would ever make. She and Sami had been planning to look for Hanna now that they were older. Maybe this timing wasn't strange at all. Raised by the Blacks to be Christians, both girls believed God worked miracles. God had given them Hanna to watch over them as small children, led them to a loving Christian home, gave Lindsey a musical gift, and a career. Sami had healed, and Emma had recovered from the flu that nearly killed her. God had been good.

Maybe another miracle, or at least a clear direction, was waiting. Maybe God was pointing her to Sal. He must have loved Hanna a lot to keep trying, even while she remains in a hospital.

Lindsey picked up the phone and called.

Sal had gone to the coffeehouse where he'd found the newspaper, hoping Lindsey might call. He didn't want to have this conversation at his parents' house, where they might eavesdrop. He was sitting near the back when his phone rang.

"Hello," Sal said.

"Hi, Sal. This is Lindsey, Hanna's sister."

Sal felt the room tilt. "Hi, Lindsey. Thank you for calling me back."

"Where are you? I think we should do this in person."

"I'm at Steam & Things on 5th and Main5th and Main."

"I'll be there in about ten minutes."

Lindsey hung up, grabbed her bike, and rode to Steam & Things on 5th and Main. She walked in and saw a young man sitting alone near the front.

"Sal?" she asked, out of breath.

"Yes. Lindsey?"

"Yes. I'm Lindsey."

"How old are you now?" he asked.

"Fourteen."

"Wow. Time's flown."

"I guess you're wondering what this is about," Sal said.

"Yes."

"The last time I saw Hanna was at my mother's funeral, quite a few years ago. We were never allowed to keep in touch with her or her foster mom. Our adoptive mother was very strict and believed Hanna wouldn't be good for us, because of the abuse and the many foster homes."

"What do you mean, 'abuse'?" Sal asked. "I knew you were taken and split up, but Hanna didn't say much, only that your mother was an alcoholic and used drugs."

"Sal… she was sold for drugs. My sister Hanna was sold for drugs."

"I'm so sorry," Sal whispered. Hanna had never told him. Of course, she hadn't; she wouldn't want to worry him.

"So Hanna is in a psychiatric hospital?" Lindsey asked softly.

"Yes. The last place she lived was a group home. Lindsey, you won't like what comes next."

Lindsey braced herself.

"She was raped and beaten by the boyfriends of one of her roommates. After everything she'd already endured, and after not being able to find you and Sami, she just… crumbled. The psychiatrist said she had nothing left. I don't know much, because I'm not family. Ruth, the group home director, could only share a little."

"Do you think there were other assaults?"

"I don't know. I met Hanna when she was eleven. She was removed from the foster home where I first met her, possible abuse. We found each other again in high school and started dating, but she would never fully open up. It was like she was holding herself together so tightly that, if she let go, the pieces would fall apart. After the rape, they finally did."

"This is horrible," Lindsey said, beginning to cry. "My God, what happened to my sister? What happened to her?" She sobbed, wiping at her eyes.

Sal wanted to comfort her, but he was the one barely holding it together. How much more could he take himself?

"What she's been through, Sal," Lindsey choked. "I would have traded places with her. We had such a good adoptive home. My mom didn't want us to be 'tainted' by anything that had happened to Hanna. If she'd done something, maybe we could have saved her!"

"Don't blame your mom," Sal said gently. "Hanna is stronger than you think. She has faith in God, too."

"Well, she must have found that later, because our mom didn't take us to church or teach us about God," Lindsey said. "We learned because the Blacks adopted us."

"Do the doctors think she'll ever get well?"

"No," Sal said. "I waited a year, hoping. It's been five now, and I'm sitting here with you. I drove to the hospital every day, but since I wasn't family, they wouldn't tell me anything. Ruth gave me small updates, but she honored the hospital's privacy rules. She did what she believed was best for Hanna, even if it meant keeping me at arm's length. Over time, going and being turned away broke me down. My parents pushed, I fell into a depression, and finally, I stopped going. Five years later, she's still there."

"How long has it been since you last saw her?"

"About five years. I last heard an update about a year ago, when I ran into Ruth, Hanna was still an inpatient. Altogether, it's been about five years since she was admitted."

"You're her sister, family. If you call, they'll talk to you. Because you're a minor, they may need to speak with your mom, but they'll talk."

"Let's call the hospital now," Lindsey said, determined. She pulled out her phone and dialed. A receptionist answered, but Lindsey missed the name of the facility.

"Is this McDermott Psychiatric Hospital?" she asked, steadying her voice.

"Yes, ma'am," the receptionist said.

"I'm looking for Dr. Richmond. Is she in?"

"Dr. Richmond is out. She'll be back later this afternoon or tomorrow morning. Would you like to leave a message?"

"Yes, please." Lindsey waited while the receptionist handled multiple lines, then dictated: "Please tell Dr. Richmond that I'm Hanna's sister. I'm only fourteen, so if she needs to speak with my parents, Emma, and Canby Black, that can be arranged. Please have her call us."

Lindsey ended the call and turned to Sal. "Let's go."

They stepped out onto the sidewalk and crossed to the parking garage. "Sal, thank you," Lindsey said in a small, breathy voice as they stopped at her locked bike.

Sal nodded sadly.

"You loved her?" she asked.

"I still do," he said.

"We need to hear from the doctor, how bad it is, whether she can understand, whether she'll recognize Sami and me."

"She'll tell you what you need to know," Sal said. "I've done what I believe God asked me to do: follow the clue in the newspaper and find you. Hanna would want you to know about her, and she would want to see you if she's coherent enough. That, I don't know. But now that I've done this, I feel lighter. I'll leave the rest with you, my friend. Hanna and I won't be again. I wish your family hope that your memories can become reality."

He leaned in and hugged Lindsey. "You're a great violinist," he said with a crooked smile.

Lindsey gave him a soft smile back. "I'll see you," she said, nodding.

She unlocked her bike and pedaled away while Sal pulled out his phone to call a friend for a ride. "Man, I need a car while I'm here," he muttered.

CHAPTER: 38

Tennis Lessons, Returning Light

"**H**anna, you are quite the tennis player," said Ruth. Hanna was hitting balls against the backboard, quick and precise.

"Are you ready to play yet, Ruth?"

"Let's do it!" Hanna said.

"I sure am," Ruth replied.

They played for a few hours. The muscle memory came back immediately for Hanna.

"I can't believe you remember how to play so well," Ruth said.

"It really isn't hard, Ruth, come on!" Hanna giggled, sliding her racquet into its case.

Ruth knew Hanna had always been good at everything she tried. Her trauma had never kept her down. She was a talented young woman, but the constant moves between foster homes had made it hard to settle long enough to discover a true passion.

So many things were coming back to her now that she was back at the group home. Ruth was proud of her. Hanna looked healthy, jogged daily, and took art classes at the library. She had even registered for college. She continued regular appointments with Dr. Richmond, and, according to the doctor, she was making remarkable progress.

Hanna and Ruth pulled up in front of the group home. Before lunch, Hanna asked for a quick stop.

Hanna jumped from the car and ran to the door. She turned and called, "Ruth, I just need to check on the girls, I'll be right back, and then we can go."

Ruth waited a few minutes. Hanna opened the front door, checked inside, locked up, and jogged back, a bit breathless.

"Sorry, Ruth. We can go now."

Ruth smiled, pulled away from the curb, and glanced over. "Hanna, why did you need to check on the girls? They're all at work at this time of day. You know the requirement to live at my home, hun."

Hanna looked puzzled. "Not the girls at the house, silly, my sisters."

Ruth kept her expression steady. Hanna's paper dolls. Recovery was moving at a stellar rate, but every time Ruth thought they were past a milestone, Hanna would backslide a little.

When Ruth asked Dr. Richmond about the backsliding, the doctor always said the paper dolls might be the last thing to go. Then she would laugh and add, "Maybe we all need paper dolls to keep ourselves undefiled." Dr. Richmond was one hundred percent satisfied with Hanna's progress. If she needed the dolls to carry on, so be it. One day she might not need them anymore and would stop. There was hope that, eventually, Hanna would not act as if they were alive in front of others, but for now, it was fine. "It's a work in progress," the doctor would say.

Ruth decided to play along. "Oh, Hanna, my mistake. How are your sisters doing?"

"They're great, but they want new clothes, so I have to make them some."

"Great," Ruth said. "And how are your other friends doing?"

"What friends?"

Okay, Ruth thought. Only the two 'sisters' today, maybe that's a good sign.

"I don't have any other friends but my sisters, Ruth, you know that," Hanna said, annoyed.

"I'm sorry, Hanna. I thought you introduced me to some friends when you first came home from the hospital. You're doing so well; I must be wrong. I apologize, honey."

"Well, Ruth, don't make things worse than they are. I have two sisters, and they are alive and well, and needing clothes I'll make after lunch. So let's go eat. I forgive you, Ruth."

CHAPTER: 39

Steam & Things, The Near Miss

Hanna and Ruth drove to their favorite coffee shop–restaurant, Steam and Things, which also happened to be Sal's favorite hangout. Sal was sitting in the back room laughing with friends, his back to the door. After Lindsey left, he'd decided to stay awhile. He did not see Hanna and Ruth.

Hanna ordered a tuna sandwich with extra pickles and a vanilla shake. Ruth ordered her usual tomato soup and a ham sandwich on rye bread.

The two of them collected their lunches and sat at a table near the front windows. "This place has the best tuna sandwiches!" said Hanna.

"Hanna, how long have you been coming here?" said Ruth. "Do you remember?"

"I remember coming with someone, but I don't remember who, Ruth. I remember the smell of chocolate and pastrami. Isn't that weird?"

Ruth knew it had been Sal, because everything had been about Sal back then. She kept eating quietly, not looking up; she didn't want to push with more questions.

In the back room, Sal kept chatting with his friends and decided to call his dad to pick him up, since his friends were having beers. He couldn't get phone service, so he started toward the front of the restaurant but stopped at the bathroom. He tossed his pastrami-sandwich wrapper and his chocolate-malt cup into the trash outside the bathroom.

"Hey, Ruth, do you care if we take our sandwiches over to the park across the street?" said Hanna. "It's such a beautiful day, and I hate to spend it inside."

"Of course, Hanna." The two of them gathered their food and drinks, walked out the front door to Ruth's car, got in, and drove to the park across the street.

Sal finished in the bathroom and went to wait for his dad at a table near the front, where he could see the curb. He sat down at the same table Ruth and Hanna had used. Suddenly, he caught a faint mix of vanilla and tuna. "Jesus. What's wrong with me? She's gone, now I'm imagining her?" He shook his head. Just then, Sal's dad pulled up. As Sal headed for the door, the clerk behind the counter called, "Hey, Sal, I think I saw Hanna a little while ago!" Sal heard the clerk say something, but couldn't make it out. He waved, walked outside, and got into his dad's car.

Meanwhile, Lindsey paced as she dialed Sami's number.

"Hello?" Sami answered.

"Hi, Sami. How are you?"

"Hey, sister! I'm great. I'm just trying to explain to Macie that teeth do not clean themselves." Lindsey laughed.

"Sami, I need to talk to you about Hanna."

"What?"

"I need to tell you something." Lindsey began to cry.

"Lindsey, what is it? Macie, go get your play-dough and bring it to the table so you can play here." Macie ran to get the play-dough and brought it to the table.

Sami took the phone into the dining room. "What, Lindsey? What's wrong?"

"It's Hanna. I found her. She's alive and,"

"Oh my God, Lindsey! Where is she? Where does she live? How did you find her? Mom and I were just making a list of possible contacts this morning."

"Sami, stop! She's sick, Sami. She's very, very sick."

"With what, Lindsey? Jesus, tell me what you're trying to tell me. Where is my sister?"

"Sami, I want to meet with you so I can tell you in person."

"What? What the hell? You call me to say this and leave out the most important information? That sucks, Lindsey! That really sucks! She's my sister, too. Tell me right now!"

"I can't, Sami. It's bad, and I need to see you in person. I'll be home as soon as I can."

"Okay," said Sami. "I don't like this. I won't sleep. I need to tell Mom what's going on."

"No, Sami," said Lindsey. "This is the best way; the only way I can do it. Goodbye. Mom and I will see you tomorrow." Lindsey hung up.

Sami was a mess. She started to cry, but didn't want Macie to see her. "I'll be right back, Macie." She went into the bathroom and cried until she could settle down. "My sister is alive and well, maybe this is a surprise, and she'll meet me at the door tomorrow." She knew she was fooling herself; Lindsey's tone had been too purposeful. Something was very, very wrong.

"Macie, come here. I need a hug." Macie came running and threw her arms around Sami. "Aww. This is why we're adopting you," Sami thought, hugging her tightly.

Macie had been visiting for a few months, and Emma and Canby had decided, without a doubt, to proceed with the adoption. The process would start immediately, and Macie would be adopted in a few months. The judge wanted the adoption expedited because she was getting older and needed to bond with a family.

The Blacks felt strongly about skipping an adoption party; Macie was still attached to her former foster parents, and they didn't want to add any more stress.

Lindsey promised to teach Macie violin, and Emma and Canby knew that learning from Lindsey meant learning from the best of the best. Sami wanted to teach her gymnastics and get her started with her coach as soon as possible.

Emma had taken Lindsey to replace the strings on her violin. Sirgi, the owner of the violin store, offered to drop Lindsey off at home so Emma could start unpacking.

"Where is my strength, Canby? I pray for strength, and I just feel so lost right now." Emma confided in Canby as she unpacked.

Canby sat in silence, listening. He knew she would love Macie just as she loved their girls. "We are older, and I have to stop pretending we aren't. I know we can do this. I'm scared, but it will be okay, we have each other."

"Let's talk about the adoption, honey. I need to feel happy right now." Canby put his arms around her, and they talked about plans for swing sets, violin lessons, and shelves in Macie's room.

Macie already had a room, even though the adoption wasn't final. She had so many toys, and the theme was princesses. The light strawberry-blonde hair that matched the girls when they were little made Emma feel she already belonged.

"It's a beautiful morning, Emma. Let's go for a jog while we wait for Lindsey," said Canby. "Great," said Emma. They told Sami they were heading out, and out the door they went.

Emma was trying to stay healthy, and Canby helped by exercising with her. Sami watched Macie. Macie played with the cat, running with a string tied to a bell so the cat would chase her. The cat's name was Lulu Bell.

He walked straight to the kitchen and started cooking pancakes as fast as he could. "Lindsey isn't here yet? Let's eat pancakes!"

Emma put two big pancakes on Macie's plate. "I want some soup," said Macie. "Syrup," said Canby. "I want a lot of sup, okay?"

Sami walked into the kitchen, looking worried. "Mom, we need to pick up the house. Lindsey will be here."

"No, now. Let's pick it up now." Sami had tears in her eyes.

They hadn't seen her this upset since the night she was removed from her birth home and placed with them. They knew the girls' history, and it hurt that the three sisters had been split and that they had played a role in keeping them apart.

Knowing the conversation with Lindsey would be hard on Sami, Canby said he'd take Macie out to lunch. Emma agreed.

Lindsey left the violin store feeling as low as she ever had. "Sami isn't going to take this well. How will we handle it? Dad will know the legalities to get Hanna out of the hospital and either bring her home or move her closer so we can visit."

It had been a long morning and was edging toward lunchtime. Canby had taken Macie out, so the house was quiet. Sami sat on the couch, staring out the window, remembering when she was a toddler.

She barely remembered anything about Hanna, except that Hanna took care of her in place of their mother. She remembered being sick and on a feeding tube. She remembered Hanna's goodness and kindness, and she wished with all her heart that Hanna had been her mother.

Lindsey remembered more than Sami did, and, over the years, she had told the stories. Lindsey, eighteen months older, remembered Hanna being carried from her bed into their mother's room and the quiet crying afterward. Hanna had handled everything and protected them. The sisters knew Hanna would never have stopped looking for them, unless something horrible had happened. And, finally, it had.

A car pulled up. Emma and Lindsey met Sami at the door. "I'm sorry, Mom, but Sami and I need to talk this through by ourselves." "I understand, girls," Emma said, heading to her bedroom.

Sami motioned for Lindsey to sit at the kitchen table, then stood. "I need some water. Do you want a bottle?" Lindsey shook her head and walked into the living room. Sami brought two bottles and sat on the couch beside her.

"No. For this, we need to be comfortable," Lindsey said with a faint smile.

Lindsey explained what Sal had told her: he had been Hanna's boyfriend and had loved her very much. Lindsey told Sami about the rape, the psychiatric hospital, and Ruth. Sami sobbed on her shoulder.

"Sami, look at me. Mom and Dad can help. They've agreed. We can get her out of there and move closer."

"We can't take care of her, Lindsey!"

"I know, but we can have her nearby. Give Mom and Dad the name of the hospital. I need the doctor's name."

"Let's do this together," said Lindsey. "We need Mom and Dad; they're the adults. We should talk to Ruth together."

"Okay," said Sami. "I want Mom to call the hospital right now. Why didn't we get this from Sal?"

"Sami, Sal is gone. He said he did what he thought was right by finding me, and he did. He followed his heart and gut, and he found me. He mentioned McDermott Hospital, I think."